Duality for Actions and Coactions of Measured Groupoids on von Neumann Algebras

Recent Titles in This Series

(Continued in the back of this publication)

MEMOIRS
of the
American Mathematical Society

Number 484

Duality for Actions and Coactions of Measured Groupoids on von Neumann Algebras

Takehiko Yamanouchi

January 1993 • Volume 101 • Number 484 (third of 4 numbers) • ISSN 0065-9266

American Mathematical Society
Providence, Rhode Island

1991 *Mathematics Subject Classification.*
Primary 46L10, 46L55; Secondary 22D25, 22D35.

Library of Congress Cataloging-in-Publication Data

Yamanouchi, Takehiko, 1961–
Duality for actions and coactions of measured groupoids on von Neumann algebras/Takehiko Yamanouchi.
p. cm. – (Memoirs of the American Mathematical Society; no. 484)
Includes bibliographical references.
ISBN 0-8218-2545-3
1. Von Neumann algebras. 2. Groupoids. 3. Duality theory (Mathematics) I. Title. II. Series.
QA3.A57 no. 484
[QA326]
510 s–dc20
[512′.55]
92-33857
CIP

Memoirs of the American Mathematical Society

This journal is devoted entirely to research in pure and applied mathematics.

Subscription information. The 1993 subscription begins with Number 482 and consists of six mailings, each containing one or more numbers. Subscription prices for 1993 are $336 list, $269 institutional member. A late charge of 10% of the subscription price will be imposed on orders received from nonmembers after January 1 of the subscription year. Subscribers outside the United States and India must pay a postage surcharge of $25; subscribers in India must pay a postage surcharge of $43. Expedited delivery to destinations in North America $30; elsewhere $92. Each number may be ordered separately; *please specify number* when ordering an individual number. For prices and titles of recently released numbers, see the New Publications sections of the *Notices of the American Mathematical Society.*

Back number information. For back issues see the *AMS Catalog of Publications.*

Subscriptions and orders should be addressed to the American Mathematical Society, P. O. Box 1571, Annex Station, Providence, RI 02901-1571. *All orders must be accompanied by payment.* Other correspondence should be addressed to Box 6248, Providence, RI 02940-6248.

Memoirs of the American Mathematical Society is published bimonthly (each volume consisting usually of more than one number) by the American Mathematical Society at 201 Charles Street, Providence, RI 02904-2213. Second-class postage paid at Providence, Rhode Island. Postmaster: Send address changes to Memoirs, American Mathematical Society, P. O. Box 6248, Providence, RI 02940-6248.

Printed in the United States of America.
This volume was printed directly from author-prepared copy.
The paper used in this book is acid-free and falls within the guidelines
established to ensure permanence and durability. ♾

10 9 8 7 6 5 4 3 2 1 98 97 96 95 94 93

Table of Contents

Abstract

Through classification of compact abelian group actions on semifinite injective factors, Jones and Takesaki introduced a notion of an action of a measured groupoid on a von Neumann algebra, which was proven to be an important tool for such an analysis. In this paper, elaborating their definition, the author introduces a new concept of a measured groupoid action that may fit more perfectly in the groupoid setting. The author also considers a notion of a coaction of a measured groupoid by showing the existence of a canonical "coproduct" on every groupoid von Neumann algebra. The construction of a crossed product by a groupoid (co)action is given. The author then proves duality between those objects, which extends Nakagami-Takesaki duality for (co)actions of locally compact groups on von Neumann algebras. Finally, various kinds of examples of groupoid (co)actions are considered.

Key words and phrases:

Duality for actions, coactions, measured groupoids, relative tensor product, coproducts, crossed product, von Neumann algebra.

§ 0. Introduction

The recent advance in the theory of operator algebras gives us a clear evidence that the analysis of the groups of automorphisms of von Neumann algebras is extremely important in order to understand the fundamental structure of von Neumann algebras. Through the analysis of those objects, many interesting, profound results have been established so far by several mathematicians, and, at the same time, plenty of important techniques have been developed. The main tool there is the construction of crossed product algebras from group actions.

If one is given a commutative group action, then Takesaki duality [**T2**] provides us with a valuable information. The most important consequence obtained by the application of this duality is the stucture theorem of von Neumann algebras of type III [**T2**].

When it comes to a noncommutative group action, since its dual object (= the set of all (equivalence classes of) irreducible representations) is no longer a group, we cannot stay within the category of groups. When we search for a suitable substitute for a dual object, we are led to consider a notion of a coaction of a group which is regarded as an action of a "group dual" (see for [**NT**] for details). The idea originates from the study of Hopf von Neumann algebras (or Kac algebras) ([**T1**], [**ES**], [**NT**]) which may be viewed as a right framework in formulating Pontryagin-van Kampen-Tannaka-Tatsuuma duality for locally compact groups. In this setting, we still have duality between those objects [**NT**] and it also gives us a structural information.

Now let us look at a special case of group actions. Namely, we consider the case in

Received by editor February 20, 1991 and in revised form October 2, 1991 which a locally compact group G is acting nonsingularly on a (Lebesgue) measure space (X, μ) (or equivalently, acting on an abelian von Neumann algebra). Then the so-called group measure space construction supplies us with a concretely realized von Neumann algebra that contains all the information necessary to analyze the given system (X, G, μ). This algebra is nothing but the crossed product algebra $L^\infty(X,\mu) \times_\alpha G$ of the derived action α of G on $L^\infty(X,\mu)$. We remark that factors of each algebraic type, type I, type II and type III can be obtained by this method. It should be also remarked that all approximately finite dimensional factors are constructed from ergodic nonsingular integer group actions (see [**HO**]). In this setting of a group action on a measure space, Dye's celebrated theorem states that what is essential in the analysis of such an algebra is to look at the groupoid (= the orbit structure) associated with the action of G on (X,μ), rather than to look at the group G itself. We note that, not only in the case of group actions on abelian algebras, groupoids naturally arise through classification of (discrete) amenable group actions on approximately finite dimensional factors, as one can see in the works of Jones-Takesaki [**JT**] and Sutherland-Takesaki [**ST**]. It should be noted too that groupoids also appear in differential geometry as holonomy groupoids naturally attached to foliated manifolds. In this direction, the noncommutative integration theory developed by A. Connes [**C1**], [**C2**] not only yields examples of operator algebras out of foliated manifolds, but also proposes a new method of studying foliated manifolds via operator algebras constructed from holonomy groupoids.

Generalizing the group measure space construction, P. Hahn obtained a systematic way of producing a von Neumann algebra $\mathcal{R}(\mathcal{G})$ from a general measured groupoid $\mathcal{G}$

[**Ha2**]. The method he obtained is "integration" of the family of "left regular representations" of a measured groupoid $\mathcal{G}$. If $\mathcal{G}$ happens to be a group, then the construction really amounts to considering the left regular representation. Thus $\mathcal{R}(\mathcal{G})$ is its group von Neumann algebra. If a measured groupoid $\mathcal{G}$ is derived from a transformation group (X, G, μ) as mentioned above, that is, $\mathcal{G} = G \times X$ with the conventional groupoid structure, then the algebra $\mathcal{R}(\mathcal{G})$ is nothing but $L^\infty(X,\mu) \times_\alpha G$. We have more structure on $\mathcal{R}(\mathcal{G})$ in this case. Namely, since $\mathcal{R}(\mathcal{G}) = L^\infty(X,\mu) \times_\alpha G$, it follows from [**NT**] that there exists the dual coaction of G on the algebra $\mathcal{R}(\mathcal{G})$. Observing these situations, one might ask himself whether it would be possible in a general setting to express the algebra $\mathcal{R}(\mathcal{G})$ as a "crossed product" of an "action" of $\mathcal{G}$ on an abelian von Neumann algebra, if he extends the crossed product construction from groups to groupoids. Another question is: "What is the 'dual coaction' on $\mathcal{R}(\mathcal{G})$ if it is possible? These questions are the very starting point of our theory of actions and coactions of groupoids on von Neumann algebras.

Now we mention the history of groupoid actions on von Neumann algebras briefly. A notion of an action of a measured groupoid on a von Neumann algebra first appeared in the work of Jones and Takesaki [**JT**]. It appeared through classification of compact abelian group actions on semifinite injective factors. Their situation is the following: Starting from an action β of a compact abelian group G on a semifinite injective factor $\mathcal{R}$, they needed to analyze the dual structure $\mathcal{M} = \mathcal{R} \times_\beta G$. For this purpose, they considered the central decomposition of $\mathcal{M}$ and then got an action of some groupoid associated with the dual action $\hat{\beta}$. In this case, von Neumann algebras appearing in the fibers in the central decomposition are all isomorphic. Hence they defined an action of a measured groupoid

$\mathcal{G}$ by a (Borel) homomorphism α of $\mathcal{G}$ into the group of all automorphisms of <u>*one*</u> von Neumann algebra. Then, using this definition, Masuda defined a crossed product of an action of a measured groupoid on a von Neumann algebra. However, he did not discuss coactions of groupoids and, therefore, he could not show any duality. Now we observe that, in general, when we consider the central decomposition of a von Neumann algebra, we do not necessarily get isomorphic algebras in the fibers. This observation suggests that, in the definition of an action of a groupoid, we should allow ourselves to have different algebras if we would like to work in the right framework. We should also keep in mind that a groupoid is a small category with inverses. So we define an action of a measured groupoid to be a functor from the groupoid into a category of von Neumann algebras with arrows as $*$-isomorphisms, satisfying some natural conditions (see § 3 for the precise definition). In order to define a notion of a coaction of a groupoid $\mathcal{G}$, we shall first show that there is a canonical "coproduct" on $\mathcal{R}(\mathcal{G})$. This morphism is a generalization of the usual coproduct on a group von Neumann algebra. Using this coproduct, we are able to define a coaction of a groupoid on a von Neumann algebra as in the case of group coactions. Then a question naturally arises whether a Takesaki duality type theorem holds in this groupoid setting. We will answer this question affirmatively. However, for duality for coactions, we have to restrict ourselves to integrable coactions in order to have much control on given coactions.

We now briefly describe the plan of this note. In § 1, we recall the notion of the relative tensor product of Hilbert spaces over von Neumann algebras due to Sauvageot [**Sa1**], [**Sa2**], [**Sa3**]. Our discussion is heavily based upon this theory. Hence we collect all the fundamental facts on the relative tensor products for the sake of completeness of our argument. Section 2 is devoted to showing the existence of a coproduct on every groupoid von Neumann algebra. To construct such a morphism, we consider a certain unitary which generalizes the fundamental unitary in the case of symmetric Kac algebras [**ES**], [**T1**]. As we mentioned above, this coproduct will play a key role when we define a coaction of a groupoid. In § 3, we give the definitions of actions and coactions of measured groupoids on von Neumann algebras. We then introduce conjugacy of two actions and coactions. We also introduce cocycle conjugacy of two actions. Section 4 is concerned with associating to a given action of a measured groupoid its coaction on a new von Neumann algebra. This algebra is called the crossed product algebra of the original action. We call the coaction on the crossed product the dual coaction. Then we shall examine how the (cocycle) conjugacy among the actions of a groupoid affects the resulting dual object. It turns out that two (cocycle) conjugate actions produce conjugate dual coactions. In § 5, we construct an action from a given coaction. The main tool in this section is the disintegrations of measures. The action obtained is called the dual action of the given coaction. Then we prove that two conjugate coactions give rise to conjugate dual actions. Section 6 is devoted to one of the main theorems proving duality for actions of measured groupoids. This really extends Takesaki duality for group actions [**T2**], [**NT**]. In § 7, we define integrability of a coaction. Then we prove duality for integrable coactions of measured groupoids. At this point, the author has been unable to show the duality in the general setting. He believes that one needs to study harmonic analysis on topological measured groupoids in order to attack the general situation. Section 8 provides several examples of actions and coactions of measured groupoids on von Neumann algebras. We shall show that every groupoid von Neumann algebra $\mathcal{R}(\mathcal{G})$ can be expressed as a crossed product of $L^\infty(X)$ by a canonical action of $\mathcal{G}$, where X stands for the unit space of the groupoid. We also give an interesting example derived from a foliated manifold. This example is essentially due to Connes [**C1**]. We should remark that all coations treated in this section are integrable ones.

This work is a revised version of the author's dissertation at University of California, Los Angeles. The author would like to express his sincere gratitude to his adviser, Professor Masamichi Takesaki, for his suggestion and encouragement.

§ 1. Relative tensor products of Hilbert spaces over abelian von Neumann algebras

Our concepts of actions and coactions of measured groupoids on von Neumann algebras are heavily based on the theory of relative tensor products of Hilbert spaces over abelian von Neumann algebras. The theory is due to Sauvageot and is thoroughly discussed in [**Sa1**], [**Sa2**], [**Sa3**] by himself. However, in order to make our argument fully self-contained, we here collect fundamental facts on the relative tensor products necessary to our discussion.

In what follows, $\mathcal{Z}$ denotes an abelian von Neumann algebra and is fixed once and for all. Let $\mathcal{M}$ be a von Neumann algbera. All the von Neumann algebras we shall consider hereafter are assumed to be the ones with separable preduals. By a Hilbert $\mathcal{M}$-module (or simply by an $\mathcal{M}$-module), we mean a (separable) Hilbert space $\mathcal{H}$ on which $\mathcal{M}$ has a nondegenerate normal representation. In most of our cases, the representations in question will be faithful ones. We write $x\xi$ for the result of the action of an element $x \in \mathcal{M}$ on a vector $\xi \in \mathcal{H}$. If $\mathcal{H}'$ is another Hilbert $\mathcal{M}$-module, then define $\mathcal{L}_{\mathcal{M}}(\mathcal{H}, \mathcal{H}')$ to be the set of all bounded operators from $\mathcal{H}$ into $\mathcal{H}'$ intertwining the two representations of $\mathcal{M}$ into $\mathcal{L}(\mathcal{H})$ and $\mathcal{L}(\mathcal{H}')$. We simply write $\mathcal{L}_{\mathcal{M}}(\mathcal{H})$ for $\mathcal{L}_{\mathcal{M}}(\mathcal{H}, \mathcal{H})$.

Suppose that $\mathcal{H}$ is as before and that φ is a faithful normal semifinite weight on $\mathcal{M}$. Then a vector $\xi \in \mathcal{H}$ is said to be φ-bounded if there exists a bounded operator $R^{\varphi}(\xi)$ from $\mathcal{H}_{\varphi}$ into $\mathcal{H}$ satisfying

$$R^{\varphi}(\xi)\eta_{\varphi}(x) = x\xi \qquad (x \in \mathcal{N}_{\varphi}),$$

where $\mathcal{N}_{\varphi}$ is a left ideal of $\mathcal{M}$ consisting of elements x with $\varphi(x^*x) < \infty$; $\mathcal{H}_{\varphi}$ is the

Hilbert space obtained by completing the pre-Hilbert space $\mathcal{N}_\varphi$ with inner product given by $(x \mid y)_\varphi = \varphi(y^*x)$; and η_φ is the canonical injection from $\mathcal{N}_\varphi$ into $\mathcal{H}_\varphi$. Let $D(\mathcal{H}, \varphi)$ denote the set of all φ-bounded vectors in $\mathcal{H}$. It is shown in **[C4]** that this set is a dense subspace of $\mathcal{H}$. It is immediate to see that $\theta^\varphi(\xi, \eta) = R^\varphi(\xi)R^\varphi(\eta)^* \quad (\xi, \eta \in D(\mathcal{H}, \varphi))$ belongs to $\mathcal{L}_\mathcal{M}(\mathcal{H})$.

Now we restrict our attention to Hilbert $\mathcal{Z}$-modules. Let $\mathcal{H}_1$ be a $\mathcal{Z}$-module, τ be a faithful normal semifinite trace on $\mathcal{Z}$, and ξ be a vector in $\mathcal{H}_1$. Then it is known that ξ is τ-bounded if and only if the Radon-Nikodym derivative $d\omega_\xi/d\tau$ of the corresponding functional ω_ξ on $\mathcal{Z}$ with respect to τ is a bounded operator, where ω_ξ stands for a vector functional given by $\omega_\xi(x) = (x\xi \mid \xi)$. Let $\mathcal{H}_2$ be another $\mathcal{Z}$-module. Then the formula

$$(\xi_1 \odot \xi_2 \mid \eta_1 \odot \eta_2) = \left(\frac{d\omega_{\xi_1,\eta_1}}{d\tau}\,\xi_2 \mid \eta_2\right)$$

defines, by extending it linearly, a sesquilinear positive form on the algebraic tensor product $D(\mathcal{H}_1, \tau) \odot \mathcal{H}_2$. Here ω_{ξ_1,η_1} is a functional defined by $\omega_{\xi_1,\eta_1}(x) = (x\xi_1 \mid \eta_1)$. We denote by $\mathcal{H}_1 \otimes_\tau \mathcal{H}_2$ the Hilbert space obtained by completing $D(\mathcal{H}_1, \tau) \odot \mathcal{H}_2$, and call it the relative tensor product of $\mathcal{Z}$-modules $\mathcal{H}_1$ and $\mathcal{H}_2$ over $\mathcal{Z}$ with respect to τ. We write $\xi_1 \otimes_\tau \xi_2$ for the image in $\mathcal{H}_1 \otimes_\tau \mathcal{H}_2$ of an elementary tensor $\xi_1 \odot \xi_2$ of $D(\mathcal{H}_1, \tau) \odot \mathcal{H}_2$. We should remark that, if one starts with the algebraic tensor product $\mathcal{H}_1 \odot D(\mathcal{H}_2, \tau)$ with a positive sesquilinear form

$$(\xi_1 \odot \xi_2 \mid \eta_1 \odot \eta_2) = \left(\frac{d\omega_{\xi_2,\eta_2}}{d\tau}\,\xi_1 \mid \eta_1\right),$$

then he arrives at the same Hilbert space $\mathcal{H}_1 \otimes_\tau \mathcal{H}_2$.

In this paragraph, we shall describe relative tensor products more concretely in terms of direct integrals of Hilbert spaces. It turns out that this concrete realization of relative tensor products is very convenient to the later discussion. Let $\mathcal{H}_1$ and $\mathcal{H}_2$ be as before. Consider a measure theoretic spectrum (X, μ) of the abelian von Neumann algebra $\mathcal{Z}$ with a faithful normal semifinite trace τ. We assume, if necessary, that X is a standard Borel space. Then, since both $\mathcal{H}_1$ and $\mathcal{H}_2$ are $\mathcal{Z}$-modules, we may consider direct integral decompositions of $\mathcal{H}_1$ and $\mathcal{H}_2$ over (X, μ):

$$\mathcal{H}_1 = \int_X^{\oplus} \mathcal{H}_1(x)\, d\mu(x) \qquad \mathcal{H}_2 = \int_X^{\oplus} \mathcal{H}_2(x)\, d\mu(x).$$

It is then easy to see that a vector $\xi = (\xi_x)_{x \in X}$ in $\mathcal{H}_1$ is τ-bounded if and only if the function $x \in X \longmapsto \|\xi_x\|$ is essentially bounded with respect to the measure μ. A field $\{\mathcal{H}_1(x) \otimes \mathcal{H}_2(x)\}_{x \in X}$ of Hilbert spaces is naturally equipped with a μ-measurable structure, so that one may form the direct integral of the measurable field over (X, μ). It can be shown that

$$\mathcal{H}_1 \otimes_\tau \mathcal{H}_2 = \int_X^{\oplus} \mathcal{H}_1(x) \otimes \mathcal{H}_2(x)\, d\mu(x).$$

Moreover, if $\xi = (\xi_x)_{x \in X} \in \mathcal{H}_1$ is τ-bounded and $\eta = (\eta_x)_{x \in X} \in \mathcal{H}_2$, then

$$\xi \otimes_\tau \eta = \int_X^{\oplus} \xi_x \otimes \eta_x\, d\mu(x).$$

This observation will become a key observation when we develop our theory of actions and coactions of measured groupoids on von Neumann algebras.

We now return to the original setting. Let x_1 and x_2 be two elements in $\mathcal{L}_{\mathcal{Z}}(\mathcal{H}_1)$ and $\mathcal{L}_{\mathcal{Z}}(\mathcal{H}_2)$, respectively. Then there exists a unique operator, denoted by $x_1 \otimes_\tau x_2$, on $\mathcal{H}_1 \otimes_\tau \mathcal{H}_2$ satisfying

$$(x_1 \otimes_\tau x_2)(\xi_1 \otimes_\tau \xi_2) = x_1 \xi_1 \otimes_\tau x_2 \xi_2 \qquad (\xi_1 \in D(\mathcal{H}_1, \tau),\ \xi_2 \in \mathcal{H}_2).$$

If z is an element in $\mathcal{Z}$, then one has

$$z \otimes_\tau 1_{\mathcal{H}_2} = 1_{\mathcal{H}_1} \otimes_\tau z.$$

This shows that the Hilbert space $\mathcal{H}_1 \otimes_\tau \mathcal{H}_2$ has a canonical $\mathcal{Z}$-module structure. There exists a unique unitary σ_τ from $\mathcal{H}_1 \otimes_\tau \mathcal{H}_2$ onto $\mathcal{H}_2 \otimes_\tau \mathcal{H}_1$ with the property that

$$\sigma_\tau(\xi_1 \otimes_\tau \xi_2) = \xi_2 \otimes_\tau \xi_1 \qquad (\xi_i \in D(\mathcal{H}_i, \tau) \ \ i = 1{,}2)$$

If x_1 and x_2 are as above, then we have

$$\sigma_\tau(x_1 \otimes_\tau x_2)\sigma_\tau{}^* = x_2 \otimes_\tau x_1$$

Through the description of relative tensor products in terms of direct integrals, this relative tensor product of operators is realized in the following way: let

$$\mathcal{H}_1 = \int_X^\oplus \mathcal{H}_1(x)\, d\mu(x), \qquad \mathcal{H}_2 = \int_X^\oplus \mathcal{H}_2(x)\, d\mu(x)$$

be as before, and let $a \in \mathcal{L}_{\mathcal{Z}}(\mathcal{H}_1)$, $b \in \mathcal{L}_{\mathcal{Z}}(\mathcal{H}_2)$. Then both a and b are decomposable operators. So let

$$a = \int_X^\oplus a(x)\, d\mu(x), \qquad b = \int_X^\oplus b(x)\, d\mu(x)$$

be their decompositions. In this case, one has

$$a \otimes_\tau b = \int_X^\oplus a(x) \otimes b(x)\, d\mu(x).$$

Let $\mathcal{H}_3$ be another Hilbert $\mathcal{Z}$-module. Since $\mathcal{H}_1 \otimes_\tau \mathcal{H}_2$ is, as we saw above, a Hilbert $\mathcal{Z}$-module again, we may form the relative tensor product $(\mathcal{H}_1 \otimes_\tau \mathcal{H}_2) \otimes_\tau \mathcal{H}_3$. At the same time, we have $\mathcal{H}_1 \otimes_\tau (\mathcal{H}_2 \otimes_\tau \mathcal{H}_3)$. However, by considering the realization of relative tensor products in terms of direct integrals, it follows that the above two spaces are equal. Thus we conclude that the operation of taking relative tensor products is associative.

If τ' is another faithful normal semifinite trace on $\mathcal{Z}$, then there exist a unique unitary joining $\mathcal{H}_1 \otimes_\tau \mathcal{H}_2$ and $\mathcal{H}_1 \otimes_{\tau'} \mathcal{H}_2$. If we look at these relative tensor products in terms of direct integrals, then we immediately notice that this unitary can be obtained from the Radon-Nikodym derivative of τ' with respect to τ. In fact, the unitary, denoted by $U_{\tau',\tau}$, from $\mathcal{H}_1 \otimes_\tau \mathcal{H}_2$ onto $\mathcal{H}_1 \otimes_{\tau'} \mathcal{H}_2$ is defined by

$$U_{\tau',\tau}(\xi_1 \otimes_\tau \xi_2) = \sqrt{\frac{d\tau'}{d\tau}}\, \xi_1 \otimes_{\tau'} \xi_2$$

whenevr $\xi_1 \in D(\mathcal{H}_1, \tau) \cap \mathcal{D}(\sqrt{\frac{d\tau'}{d\tau}})$, $\xi_2 \in \mathcal{H}_2$. $\mathcal{D}(T)$ stands for the domain of a (densely defined) unbounded operator T. If x_1 and x_2 belong to $\mathcal{L}_\mathcal{Z}(\mathcal{H}_1)$ and $\mathcal{L}_\mathcal{Z}(\mathcal{H}_2)$, respectively. then we have

$$U_{\tau',\tau}(x_1 \otimes_\tau x_2) U^*_{\tau',\tau} = x_1 \otimes_{\tau'} x_2.$$

Because of this fact, we simply write $x_1 \otimes_\mathcal{Z} x_2$ for $x_1 \otimes_\tau x_2$, if there is no danger of confusion. Suppose that $\mathcal{N}_i$, $(i = 1,2)$ is a von Neumann subalgebra of $\mathcal{L}_\mathcal{Z}(\mathcal{H}_i)$, then we denote by $\mathcal{N}_1 \otimes_\mathcal{Z} \mathcal{N}_2$ the von Neumann algebra generated by the elements of the form $x_1 \otimes_\mathcal{Z} x_2$, $(x_i \in \mathcal{N}_i,\ (i = 1,2))$.

Next we introduce a notion of a fiber product of von Neumann algebras over $\mathcal{Z}$. Let $\mathcal{H}_1$ and $\mathcal{H}_2$ be as before. Suppose that $\mathcal{M}_1$ and $\mathcal{M}_2$ are von Neumann algebras on $\mathcal{H}_1$ and $\mathcal{H}_2$, respectively, both of which are containing $\mathcal{Z}$ as a von Neumann subalgebra (we say, in this case, that $\mathcal{M}_1$ and $\mathcal{M}_2$ are $\mathcal{Z}$-modules). Then the fiber product $\mathcal{M}_1 *_\mathcal{Z} \mathcal{M}_2$ of $\mathcal{M}_1$ and $\mathcal{M}_2$ over $\mathcal{Z}$ is defined to be the commutant of $\mathcal{L}_{\mathcal{M}_1}(\mathcal{H}_1) \otimes_\mathcal{Z} \mathcal{L}_{\mathcal{M}_2}(\mathcal{H}_2)$ in $\mathcal{H}_1 \otimes_\tau \mathcal{H}_2$.

One can show that the fiber product operation is also associative. This concept will be used when we define a coaction of a measured groupoid on a von Neumann algebra.

We have the following proposition which relates fiber products over $\mathcal{Z}$ to relative tensor products over $\mathcal{Z}$ of two von Neumann algebras.

Proposition 1.1. *Suppose that* $\{\mathcal{M}_1, \mathcal{H}_1\}$ *and* $\{\mathcal{M}_2, \mathcal{H}_2\}$ *are* $\mathcal{Z}$*-modules.*

(1) *If* $\mathcal{Z}$ *is contained in the center of* $\mathcal{M}_1$ *and* $\mathcal{M}_2$*, respectively, then we have*

$$\mathcal{M}_1 *_{\mathcal{Z}} \mathcal{M}_2 = \mathcal{M}_1 \otimes_{\mathcal{Z}} \mathcal{M}_2.$$

(2) *If* $\mathcal{Z}$ *is contained only in the center of* $\mathcal{M}_1$*, then*

$$\mathcal{M}_1 *_{\mathcal{Z}} \mathcal{M}_2 = \mathcal{M}_1 \otimes_{\mathcal{Z}} (\mathcal{M}_2 \cap \mathcal{Z}').$$

As an easy consequence of (2), *we conclude that* $\mathcal{Z} *_{\mathcal{Z}} \mathcal{M}_2$ *is isomorphic to* $\mathcal{M}_2 \cap \mathcal{Z}'$ *and that* $\mathcal{Z} *_{\mathcal{Z}} \mathcal{Z}$ *is canonically isomorphic to* $\mathcal{Z}$.

(3) *The commutant of* $\mathcal{L}_{\mathcal{Z}}(\mathcal{H}_1) \otimes_{\mathcal{Z}} \mathbf{C}_{\mathcal{H}_2}$ *in* $\mathcal{H}_1 \otimes_\tau \mathcal{H}_2$ *is equal to* $\mathbf{C}_{\mathcal{H}_1} \otimes_{\mathcal{Z}} \mathcal{L}_{\mathcal{Z}}(\mathcal{H}_2)$.

(4) *The center of* $\mathcal{L}_{\mathcal{Z}}(\mathcal{H}_1) \otimes_{\mathcal{Z}} \mathcal{L}_{\mathcal{Z}}(\mathcal{H}_2)$ *is equal to* $\mathcal{Z}$.

The proof is found in [**Sa2**].

Next we discuss a notion of a fiber product of two representations over $\mathcal{Z}$. Among other facts on this fiber product of representations, the following proposition becomes a very essential result when we define a coaction of a measured groupoid on a von Neumann algebra. First we describe the situation. Let $\mathcal{H}_1$, $\mathcal{H}_2$, $\mathcal{M}_1$ and $\mathcal{M}_2$ be as before. Suppose that $\mathcal{K}_1$ and $\mathcal{K}_2$ are Hilbert spaces, and that π_1 and π_2 are normal representations of $\mathcal{M}_1$ and $\mathcal{M}_2$, respectively, into $\mathcal{L}(\mathcal{K}_1)$ and $\mathcal{L}(\mathcal{K}_2)$: then $\mathcal{K}_1$ and $\mathcal{K}_2$ are equipped with

$\mathcal{Z}$-module structures. In most of our discussion, we deal with faithful π_1 and π_2. Under this situation, we have

Proposition 1.2. (1) *There exists a unique representation, denoted by* $\pi_1 *_{\mathcal{Z}} \pi_2$, *of* $\mathcal{M}_1 *_{\mathcal{Z}} \mathcal{M}_2$ *into* $\mathcal{L}(\mathcal{K}_1 \otimes_\tau \mathcal{K}_2)$ *satisfying the following:*

$$(T_1 \otimes_{\mathcal{Z}} T_2) Y \xi = (\pi_1 *_{\mathcal{Z}} \pi_2)(Y)(T_1 \otimes_{\mathcal{Z}} T_2)\xi$$

whenever T_i *in* $\mathcal{L}_{\mathcal{M}_i}(\mathcal{H}_i, \mathcal{K}_i)$ $(i = 1, 2)$, $Y \in \mathcal{M}_1 *_{\mathcal{Z}} \mathcal{M}_2$ *and* $\xi \in \mathcal{H}_1 \otimes_\tau \mathcal{H}_2$.

(2) *We have*

$$(\pi_1 *_{\mathcal{Z}} \pi_2)(\mathcal{M}_1 *_{\mathcal{Z}} \mathcal{M}_2) = \pi_1(\mathcal{M}_1) *_{\mathcal{Z}} \pi_2(\mathcal{M}_2).$$

(3) *If both* π_1 *and* π_2 *are faithful, then so is* $\pi_1 *_{\mathcal{Z}} \pi_2$.

The proof is available in [**Sa2**].

Because of the uniqueness of the fiber product of two representations, we obtain, as a simple consequence of that, the following corollary.

Corollary 1.3. *Let* $\mathcal{M}_1$, $\mathcal{M}_2$, π_1 *and* π_2 *be as before. We put* $\mathcal{N}_i = \pi_i(\mathcal{M}_i)$ $(i = 1,2)$. $\mathcal{N}_1$ *and* $\mathcal{N}_2$ *become* $\mathcal{Z}$*-modules. Suppose that* ρ_1 *and* ρ_2 *are normal representations of* $\mathcal{N}_1$ *and* $\mathcal{N}_2$, *respectively, into* $\mathcal{L}(\mathcal{K}'_1)$ *and* $\mathcal{L}(\mathcal{K}'_2)$. *Then we have*

$$(\rho_1 *_{\mathcal{Z}} \rho_2) \circ (\pi_1 *_{\mathcal{Z}} \pi_2) = (\rho_1 \circ \pi_1) *_{\mathcal{Z}} (\rho_2 \circ \pi_2).$$

This corollary implies a functorial property of the $\mathcal{Z}$-product in the category of von Neumann algebras.

§ 2. Coproducts of groupoid von Neumann algebras

In [**Ha2**], P. Hahn obtained a way of producing a von Neumann algebra from a given measured groupoid. This method amounts to the so-called group-measure space construction when the groupoid is derived from a group action on a measure space. This section is devoted to constructing a "coproduct" of every von Neumann algebra arising from a measured groupoid by Hahn's method. If a groupoid in question happens to be a group, then the von Neumann algebra obtained by Hahn's machinery is nothing but its group algebra and our "coproduct" is the conventional coproduct (comultiplication) of the group algebra in the sense of Hopf algebra or Kac algebra [**ES**], [**T1**]. This situation justifies our terminology of "coproduct". This coproduct will play a key role when we define a coaction of a measured groupoid on a von Neumann algebra.

Throughout this note, we fix a standard Borel groupoid $\mathcal{G}$. We assume that all relevant maps and sets that are related to the groupoid structure of $\mathcal{G}$ are Borel. We denote the source (resp. the range) of an element γ of the groupoid by $s(\gamma)$ (resp. $r(\gamma)$). The unit space of $\mathcal{G}$, which is the image of the groupoid under the source (or the range) map, is denoted by X. For every $x \in X$, $\mathcal{G}^x$ (resp. $\mathcal{G}_x$) designates the inverse image of the range (resp. the source) map $r^{-1}(x)$ (resp. $s^{-1}(x)$). We assume from now on that the groupoid admits a faithful proper transverse function $\{\lambda^x\}_{x\in X}$ and a transverse measure Λ with a module δ. For these terminologies and basic properties of transverse functions and transverse measures, we refer readers to [**C2**]. Given such a system $(\{\lambda^x\}_{x\in X}, \Lambda, \delta)$ on $\mathcal{G}$, we have a σ-finite measure, denoted by Λ_λ (or, simply by μ if there is no danger of confusion), on the unit space X. Let ν be the σ-finite measure on $\mathcal{G}$ given by integrating the transverse function against the measure μ. We also let λ_x $(x \in X)$ and ν^{-1} be the images of λ^x and ν, respectively, by the inverse map. By definition, ν is equivalent to ν^{-1} in the sense of absolute continuity and the module δ is the Radon-Nikodym derivative $d\nu/d\nu^{-1}$. Hereafter we call a groupoid with such a system a measured groupoid.

For a Borel space $(Y, \mathcal{B})$, $\mathcal{F}(Y)$ (resp. $\mathcal{F}^+(Y)$) is the set of all Borel (resp. positive extended real valued Borel) functions on Y. For any $f, g \in \mathcal{F}(\mathcal{G})$, we define a convolution product $f*g$ by

$$(f*g)(\gamma) = \int f(\gamma_1)g(\gamma_1{}^{-1}\gamma)\, d\lambda^{r(\gamma)}(\gamma_1) \qquad (\gamma \in \mathcal{G}).$$

Moreover, for every f in $\mathcal{F}(\mathcal{G})$, we define $f^\sharp$, $f^\flat$ and $\check{f}$ to be

$$f^\sharp(\gamma) = \delta(\gamma)^{-1}\overline{f(\gamma^{-1})}, \qquad f^\flat(\gamma) = \overline{f(\gamma^{-1})},$$
$$\check{f}(\gamma) = f(\gamma^{-1}), \qquad (\gamma \in \mathcal{G}).$$

For f as above, let

$$\|f\|_I = \max\{\, \|\lambda(|f|)\|_\infty,\ \|\lambda(|f^\sharp|)\|_\infty \,\},$$

where $\lambda(g)$ is a function on X given by $\lambda(g)(x) = \int g(\gamma)\, d\lambda^x(\gamma)$, $(g \in \mathcal{F}(\mathcal{G}))$.

A function $f \in \mathcal{F}(\mathcal{G})$ is said to be strictly δ_a-bounded if there is a number $a \geq 1$ such that, for all $\gamma \in \mathcal{G}$, $f(\gamma) = 0$ or $\delta(\gamma) \in [1/a, a]$. A Borel function g is δ_a-bounded if g agrees a.e. with a strictly δ_a-bounded function. A function is called δ-bounded if it is δ_a-bounded for some $a \geq 1$. Then we let

$$\mathcal{A}_I = \{\, \xi \in L^2(\mathcal{G}, \nu) : \xi \text{ is } \delta\text{-bounded},\ \ \|\xi\|_I < \infty \,\}$$

It is shown in **[Ha2]** that, letting $Jf = \delta^{1/2} f^\sharp$, $(f \in \mathcal{F}(\mathcal{G}))$, $\mathcal{A}_I$ ($\mathcal{U}_I$ in his notation) with multiplication $*$, left involution $\sharp$, is a left Hilbert algebra with the corresponding modular conjugation J, the right involution $\flat$ and the modular operator Δ given by

$$\Delta\xi = \delta\xi, \qquad (\xi \in \mathcal{A}_I).$$

We write $L(\xi)$ for the left multiplication of $\xi \in \mathcal{A}_I$ and call it the regular representation of $\mathcal{G}$ with respect to the system $(\lambda, \Lambda, \delta)$. Let $\mathcal{R}(\mathcal{G}, \lambda, \Lambda, \delta)$ (or simply $\mathcal{R}(\mathcal{G})$) denote the left von Neumann algebra associated with the left Hilbert algebra $\mathcal{A}_I$. It is often called the groupoid von Neumann algebra of $\mathcal{G}$ derived from $(\lambda, \Lambda, \delta)$.

Besides the set of multiplicative pairs $\mathcal{G}^{(2)}$, we introduce three important Borel subsets $\mathcal{H}^{(2)}$, $\mathcal{F}^{(2)}$ and $\mathcal{I}^{(2)}$ of $\mathcal{G} \times \mathcal{G}$ as follows:

$$\mathcal{H}^{(2)} = \{ (\gamma_1, \gamma_2) \in \mathcal{G} \times \mathcal{G} : r(\gamma_1) = r(\gamma_2) \},$$

$$\mathcal{F}^{(2)} = \{ (\gamma_1, \gamma_2) \in \mathcal{G} \times \mathcal{G} : s(\gamma_2) = r(\gamma_1) \},$$

$$\mathcal{I}^{(2)} = \{ (\gamma_1, \gamma_2) \in \mathcal{G} \times \mathcal{G} : s(\gamma_1) = s(\gamma_2) \}.$$

There exist canonically defined measures ν_1, ν_2, ν_3 and ν_4 on these Borel subsets $\mathcal{H}^{(2)}$, $\mathcal{G}^{(2)}$, $\mathcal{F}^{(2)}$ and $\mathcal{I}^{(2)}$, respectively. For example, the measure ν_1 is determined by the integral

$$\begin{aligned}\int_{\mathcal{H}^{(2)}} f(\gamma_1, \gamma_2)\, d\nu_1(\gamma_1, \gamma_2) &= \int\int\int f(\gamma_1, \gamma_2) d\lambda^x(\gamma_2) d\lambda^x(\gamma_1) d\mu(x) \\ &= \int\int\int f(\gamma_1, \gamma_2) d\lambda^x(\gamma_1) d\lambda^x(\gamma_2) d\mu(x),\end{aligned}$$

where $f \in \mathcal{F}^+(\mathcal{H}^{(2)})$. The other three measures ν_2, ν_3 and ν_4 are given by the following equations respectively:

$$\begin{aligned}\int_{\mathcal{G}^{(2)}} f(\gamma_1, \gamma_2) d\nu_2(\gamma_1, \gamma_2) &= \int\int\int f(\gamma_1, \gamma_2) d\lambda^{s(\gamma_1)}(\gamma_2) d\lambda^x(\gamma_1) d\mu(x), \\ \int_{\mathcal{F}^{(2)}} f(\gamma_1, \gamma_2) d\nu_3(\gamma_1, \gamma_2) &= \int\int\int f(\gamma_1, \gamma_2) d\lambda^{s(\gamma_2)}(\gamma_1) d\lambda^x(\gamma_2) d\mu(x), \\ \int_{\mathcal{I}^{(2)}} f(\gamma_1, \gamma_2) d\nu_4(\gamma_1, \gamma_2) &= \int\int\int f(\gamma_1, \gamma_2) d\lambda'_x(\gamma_2) d\lambda'_x(\gamma_1) d\mu(x),\end{aligned}$$

where f is a positive Borel function on an appropriate Borel space and $d\lambda'_x = \delta d\lambda_x$.

Let $M(f)$ denote the (bounded) operator on $L^2(\mathcal{G}, \nu)$ of multiplication by a function f in $L^\infty(\mathcal{G}, \nu)$. We will write $\mathcal{Z}$ for the abelian von Neumann algebra $L^\infty(X, \mu)$. The notation will be fixed from now on. The Hilbert space $L^2(\mathcal{G}, \nu)$ can be equipped naturally with a $\mathcal{Z}$-bimodule structure in the following manner: the left action of $\mathcal{Z}$ is a representation sending $h \in \mathcal{Z}$ to $M(h \circ r)$: and the right action of $\mathcal{Z}$ on $L^2(\mathcal{G}, \nu)$ is then given by the map carrying $h \in \mathcal{Z}$ to $M(h \circ s)$. As usual, we write $h\xi$ for $M(h \circ r)\xi$, $(\xi \in L^2(\mathcal{G}, \nu)$. On the other hand, we denote by ξh the vector $M(h \circ s)\xi$. Set

$$\mathcal{Z}_R = \{\, M(h \circ r) : h \in \mathcal{Z} \,\},$$

$$\mathcal{Z}_S = \{\, M(h \circ s) : h \in \mathcal{Z} \,\}.$$

Note that, due to Theorem 3.4 of [**Ha2**], $\mathcal{Z}_R$ is a von Neumann subalgebra of the groupoid von Neumann algebra $\mathcal{R}(\mathcal{G})$. In fact, we have that $L(h\xi) = M(h \circ r)L(\xi)$ for any $h \in \mathcal{Z}$ and any $\xi \in \mathcal{A}_I$. Nondegeneracy of the regular representation L then yields the above claim. The right action of $\mathcal{Z}$ in turn commutes with $\mathcal{R}(\mathcal{G})$. We write ${}_{\mathcal{Z}}L^2(\mathcal{G}, \nu)$ (resp. $L^2(\mathcal{G}, \nu)_{\mathcal{Z}}$) when the left (resp. right) action is specifically considered on $L^2(\mathcal{G}, \nu)$. According to the two actions of $\mathcal{Z}$ on $L^2(\mathcal{G}, \nu)$ described above, $L^2(\mathcal{G}, \nu)$ is decomposed into:

$$ {}_{\mathcal{Z}}L^2(\mathcal{G}, \nu) = \int_X^\oplus L^2(\mathcal{G}^x, \lambda^x) d\mu(x),$$

$$ L^2(\mathcal{G}, \nu)_{\mathcal{Z}} = \int_X^\oplus L^2(\mathcal{G}_x, \lambda'_x) d\mu(x).$$

We will freely identify every square integralable section $\{\xi_x\}_{x \in X}$ of the measurable field $\{L^2(\mathcal{G}^x, \lambda^x)\}_{x \in X}$ (resp. $\{L^2(\mathcal{G}_x, \lambda'_x)\}_{x \in X}$) with a ν-square integrable function ξ on $\mathcal{G}$ by $\xi(\gamma) = \xi_x(\gamma)$ for $\gamma \in \mathcal{G}^x$ (resp. $\gamma \in \mathcal{G}_x$) and $x \in X$. Then, from the fact we noted in the previous section, we obtain the following lemma.

Lemma 2.1. *With the notation as above, we have*

$$D({}_{\mathcal{Z}}L^2(\mathcal{G},\nu),\mu) = \{\, \xi \in L^2(\mathcal{G},\nu) : \lambda(|\xi|^2) \in L^\infty(X,\mu) \,\},$$

$$D(L^2(\mathcal{G},\nu)_{\mathcal{Z}},\mu) = \{\, \xi \in L^2(\mathcal{G},\nu) : \lambda'(|\xi|^2) \in L^\infty(X,\mu) \,\},$$

where $\lambda'(f)$ *is a function on* X *defined by* $\lambda'(f)(x) = \int f(\gamma)d\lambda'_x(\gamma) \quad (f \in \mathcal{F}(\mathcal{G}))$.

If $f, g \in D({}_{\mathcal{Z}}L^2(\mathcal{G},\nu),\mu)$, then, from the discussion in § 1, the Radon-Nikodym derivative $d\omega_{f,g}/d\mu$ falls in $\mathcal{Z}$. Put $< f,\, g >= d\omega_{f,g}/d\mu$. Then it is a simple calculation to verify that

$$\begin{aligned} < f,\, g > (x) &= \lambda(f\,\overline{g})(x) \\ &= \int f(\gamma)\overline{g(\gamma)}\, d\lambda^x(\gamma). \end{aligned}$$

If, in turn, f, g are in $D(L^2(\mathcal{G},\nu)_{\mathcal{Z}},\mu)$, then the Radon-Nikodym derivative $< f,\, g >^\circ = d\omega_{f,g}/d\mu$ is given by

$$\begin{aligned} < f,\, g >^\circ (x) &= \lambda'(f\,\overline{g})(x) \\ &= \int f(\gamma)\overline{g(\gamma)}\, d\lambda'_x(\gamma). \end{aligned}$$

This observation enables us to describe the structure of a relative tensor product of $L^2(\mathcal{G},\nu)_{\mathcal{Z}} \otimes_\mu {}_{\mathcal{Z}}L^2(\mathcal{G},\nu)$ concretely. In fact, if $f_1, g_1 \in L^2(\mathcal{G},\nu)$ and $f_2, g_2 \in D({}_{\mathcal{Z}}L^2(\mathcal{G},\nu),\mu)$, then, setting $f = f_1 \otimes_\mu f_2,\ g = g_1 \otimes_\mu g_2$, we have

$$\begin{aligned} (f \mid g) &= (f_1 < f_2,\, g_2 > \mid g_1) \\ &= \int f_1(\gamma_1) < f_2,\, g_2 > (s(\gamma_1))\overline{g_1(\gamma_1)}\, d\nu(\gamma_1) \\ &= \int\int\int f_1(\gamma_1) f_2(\gamma_2)\overline{g_1(\gamma_1) g_2(\gamma_2)} d\lambda^{s(\gamma_1)}(\gamma_2) d\lambda^x(\gamma_1) d\mu(x). \end{aligned}$$

From this calculation, it easily follows that $L^2(\mathcal{G},\nu)_{\mathcal{Z}} \otimes_\mu {}_{\mathcal{Z}}L^2(\mathcal{G},\nu)$ can be identified with the Hilbert space $L^2(\mathcal{G}^{(2)},\nu_2)$ via the map : $f \otimes_\mu g \in L^2(\mathcal{G},\nu)_{\mathcal{Z}} \otimes_\mu {}_{\mathcal{Z}}L^2(\mathcal{G},\nu) \longmapsto f \times g|_{\mathcal{G}^{(2)}}$, where $f \times g$ is a function on $\mathcal{G} \times \mathcal{G}$ defined by $(f \times g)(\gamma_1, \gamma_2) = f(\gamma_1)g(\gamma_2)$. By a similar computation as above, we may conclude the following proposition.

Proposition 2.2. *There are natural identifications among the following Hilbert spaces.*

$$ {}_{Z}L^2(\mathcal{G},\nu)\otimes_{\mu Z}L^2(\mathcal{G},\nu) \cong L^2(\mathcal{H}^{(2)},\nu_1) $$

$$ {}_{Z}L^2(\mathcal{G},\nu)\otimes_{\mu}L^2(\mathcal{G},\nu)_Z \cong L^2(\mathcal{F}^{(2)},\nu_3) $$

$$ L^2(\mathcal{G},\nu)_Z\otimes_{\mu}L^2(\mathcal{G},\nu)_Z \cong L^2(\mathcal{I}^{(2)},\nu_4). $$

From now on, we shall freely identify these Hilbert spaces without referring to the unitaries joining isomorphic spaces.

We construct below three unitaries among the above Hilbert spaces, which play an important role in the later discussion. These operators really reflect the groupoid structure of $\mathcal{G}$. We shall define operators $U : L^2(\mathcal{I}^{(2)},\nu_4) \longrightarrow L^2(\mathcal{G}^{(2)},\nu_2)$, $V : L^2(\mathcal{H}^{(2)},\nu_1) \longrightarrow L^2(\mathcal{F}^{(2)},\nu_3)$ and $W : L^2(\mathcal{H}^{(2)},\nu_1) \longrightarrow L^2(\mathcal{G}^{(2)},\nu_2)$ respectively by

$$ \{U\zeta\}(\gamma_1,\gamma_2) = \delta(\gamma_2)^{1/2}\zeta(\gamma_1\gamma_2,\gamma_2) \qquad ((\gamma_1,\gamma_2) \in \mathcal{G}^{(2)}) $$

$$ \{V\xi\}(\gamma_1,\gamma_2) = \xi(\gamma_2\gamma_1,\gamma_2) \qquad ((\gamma_1,\gamma_2) \in \mathcal{F}^{(2)}) $$

$$ \{W\xi\}(\gamma_1,\gamma_2) = \xi(\gamma_1,\gamma_1\gamma_2) \qquad ((\gamma_1,\gamma_2) \in \mathcal{G}^{(2)}), $$

where $\zeta \in L^2(\mathcal{I}^{(2)},\nu_4)$ and $\xi \in L^2(\mathcal{H}^{(2)},\nu_1)$. In the case of $\mathcal{G}$ being a locally compact group G, the operator W is called the fundamental unitary operator of G (refer to [**T1**]).

Proposition 2.3. *The operators U, V and W are unitaries whose inverses are given respectively by*

(i) $\{U^*\eta'\}(\gamma_1,\gamma_2) = \delta(\gamma_2)^{-1/2}\eta'(\gamma_1{\gamma_2}^{-1},\gamma_2)$

(ii) $\{V^*\xi\}(\gamma_1,\gamma_2) = \xi({\gamma_2}^{-1}\gamma_1,\gamma_2)$

(iii) $\{W^*\eta'\}(\gamma_1,\gamma_2) = \eta'(\gamma_1,{\gamma_1}^{-1}\gamma_2),$

where $\eta' \in L^2(\mathcal{G}^{(2)}, \nu_2)$ *and* $\xi \in L^2(\mathcal{F}^{(2)}, \nu_3)$.

Proof. We will treat only the operator U here. The proof for the other two operators are basically the same as the one presented below.

We have

$$\begin{aligned}
&\int |\delta(\gamma_2)^{1/2}\eta(\gamma_1\gamma_2, \gamma_2)|^2 d\nu_2(\gamma_1, \gamma_2) \\
&= \int\int\int \delta(\gamma_2)|\eta(\gamma_1\gamma_2, \gamma_2)|^2 d\lambda^{s(\gamma_1)}(\gamma_2) d\lambda^x(\gamma_1) d\mu(x) \\
&= \int\int\int |\eta(\gamma_1\gamma_2, \gamma_2)|^2 \delta(\gamma_2) d\lambda^{s(\gamma_1)}(\gamma_2)\delta(\gamma_1) d\lambda_x(\gamma_1) d\mu(x) \\
&= \int\int\int |\eta(\gamma_1\gamma_2, \gamma_2)|^2 \delta(\gamma_1\gamma_2) d\lambda_{r(\gamma_2)}(\gamma_1) d\lambda^x(\gamma_2) d\mu(x) \\
&= \int\int\int |\eta(\gamma_1, \gamma_2)|^2 \delta(\gamma_1) d\lambda_{s(\gamma_2)}(\gamma_1)\delta(\gamma_2) d\lambda_x(\gamma_2) d\mu(x) \\
&= \int\int\int |\eta(\gamma_1, \gamma_2)|^2 d\lambda'_x(\gamma_2) d\lambda'_x(\gamma_1) d\mu(x) \\
&= \|\eta\|^2.
\end{aligned}$$

The third and sixth equalities are guaranteed by Fubini's thoerem. This computation shows that U is an isometry. Moreover, we have

$$\begin{aligned}
&\int\int\int |\delta(\gamma_2)^{-1/2}\eta'(\gamma_1\gamma_2{}^{-1}, \gamma_2)|^2 d\lambda'_x(\gamma_2) d\lambda'_x(\gamma_1) d\mu(x) \\
&= \int\int\int |\eta'(\gamma_1\gamma_2{}^{-1}, \gamma_2)|^2 \delta(\gamma_1) d\lambda_x(\gamma_2) d\lambda_x(\gamma_1) d\mu(x) \\
&= \int\int\int |\eta'(\gamma_1\gamma_2{}^{-1}, \gamma_2)|^2 \delta(\gamma_1) d\lambda_{s(\gamma_2)}(\gamma_1) d\lambda_x(\gamma_2) d\mu(x) \\
&= \int\int\int |\eta'(\gamma_1, \gamma_2)|^2 \delta(\gamma_1\gamma_2) d\lambda_{r(\gamma_2)}(\gamma_1) d\lambda_x(\gamma_2) d\mu(x) \\
&= \int\int\int |\eta'(\gamma_1, \gamma_2)|^2 \delta(\gamma_1) d\lambda_{r(\gamma_2)}(\gamma_1) d\lambda^x(\gamma_2) d\mu(x) \\
&= \int\int\int |\eta'(\gamma_1, \gamma_2)|^2 d\lambda^{s(\gamma_1)}(\gamma_2)\delta(\gamma_1) d\lambda_x(\gamma_1) d\mu(x) \\
&= \int\int\int |\eta'(\gamma_1, \gamma_2)|^2 d\lambda^{s(\gamma_1)}(\gamma_2) d\lambda^x(\gamma_1) d\mu(x) \\
&= \|\eta'\|^2.
\end{aligned}$$

The second and fifth equalities are justified again by Fubini's theorem. This shows that equation (i) indeed defines an isometry, which can be easily proven to be the inverse of U. Therefore, U is a unitary. Q.E.D.

According to the discussion in § 1, there exists a unique unitary $\sigma_{1,\mu}$ (resp. $\sigma_{2,\mu}$) from $L^2(\mathcal{G},\nu)_Z \otimes_\mu {}_Z L^2(\mathcal{G},\nu)$ onto ${}_Z L^2(\mathcal{G},\nu) \otimes_\mu L^2(\mathcal{G},\nu)_Z$ (resp. ${}_Z L^2(\mathcal{G},\nu) \otimes_\mu {}_Z L^2(\mathcal{G},\nu)$ onto itself) satisfying

$$\sigma_{1,\mu}(\xi_1 \otimes_\mu \xi_2) = \xi_2 \otimes_\mu \xi_1$$

$$(resp.\quad \sigma_{2,\mu}(\eta_1 \otimes_\mu \eta_2) = \eta_2 \otimes_\mu \eta_1\,)$$

where the vector ξ_1 belongs to the set $D(L^2(\mathcal{G},\nu)_Z,\mu)$, and ξ_2 lies in $D({}_Z L^2(\mathcal{G},\nu),\mu)$ (resp. $\eta_i \in D({}_Z L^2(\mathcal{G},\nu),\mu)$ $i = 1,2$). In terms of the identifications obtained in Proposition 2.2 and from the argument precedeing it, $\sigma_{1,\mu}$ and $\sigma_{2,\mu}$ can be described more explicitly in the following way:

$$\{\sigma_{1,\mu}\xi\}(\gamma_1,\gamma_2) = \xi(\gamma_2,\gamma_1) \qquad ((\gamma_1,\gamma_2) \in \mathcal{F}^{(2)}\,)$$

$$\{\sigma_{2,\mu}\eta\}(\gamma_1,\gamma_2) = \eta(\gamma_2,\gamma_1) \qquad ((\gamma_1,\gamma_2) \in \mathcal{H}^{(2)}\,),$$

where $\xi \in L^2(\mathcal{G}^{(2)},\nu_2)$ and $\eta \in L^2(\mathcal{H}^{(2)},\nu_1)$. Then we obtain

Lemma 2.4. *With the notation as above, we have*

$$V = \sigma_{1,\mu} W \sigma_{2,\mu}$$

Proof. If $\eta \in L^2(\mathcal{H}^{(2)},\nu_1)$ and $(\gamma_1,\gamma_2) \in \mathcal{F}^{(2)}$, then

$$\begin{aligned}\{\sigma_{1,\mu} W \sigma_{2,\mu}\eta\}(\gamma_1,\gamma_2) &= \{W\sigma_{2,\mu}\eta\}(\gamma_2,\gamma_1)\\ &= \{\sigma_{2,\mu}\eta\}(\gamma_2,\gamma_2\gamma_1)\\ &= \eta(\gamma_2\gamma_1,\gamma_2)\\ &= \{V\eta\}(\gamma_1,\gamma_2).\end{aligned}$$

This proves the lemma. Q.E.D.

As we noted before, the right action of $\mathcal{Z}$ on $L^2(\mathcal{G},\nu)$ commutes with $\mathcal{R}(\mathcal{G})$. Hence, on the Hilbert space ${}_{\mathcal{Z}}L^2(\mathcal{G},\nu)\otimes_\mu L^2(\mathcal{G},\nu)_{\mathcal{Z}}$, it does have a perfect meaning to form a relative tensor product $1\otimes_{\mathcal{Z}} a$ over $\mathcal{Z}$ of the identity operator 1 on $L^2(\mathcal{G},\nu)$ and any member a of $\mathcal{R}(\mathcal{G})$. Accordingly, we may define a $*$-isomorphism Γ from $\mathcal{R}(\mathcal{G})$ into the set of all bounded operators $\mathcal{L}({}_{\mathcal{Z}}L^2(\mathcal{G},\nu)\otimes_{\mu\mathcal{Z}}L^2(\mathcal{G},\nu))$ by

$$\Gamma(a) = V^*(1\otimes_{\mathcal{Z}} a)V \qquad (a \in \mathcal{R}(\mathcal{G})).$$

In what follows, we shall give a concrete expression of $\Gamma(a)$ when a is a typical element in $\mathcal{R}(\mathcal{G})$ of the form $a = L(f)$ for some $f \in \mathcal{A}_I$. Using this expression, we will be able to describe the image of Γ in more detailed manner and then prove that the morphism Γ satisfies a "coassociative" property.

Let $f \in \mathcal{A}_I$, $\xi \in {}_{\mathcal{Z}}L^2(\mathcal{G},\nu)\otimes_{\mu\mathcal{Z}}L^2(\mathcal{G},\nu) \cong L^2(\mathcal{H}^{(2)},\nu_1)$ and $(\gamma_1,\gamma_2) \in \mathcal{H}^{(2)}$. Then we calculate

$$\begin{aligned}
\{V^*(1\otimes_{\mathcal{Z}} L(f))V\xi\}(\gamma_1,\gamma_2) &= \{(1\otimes_{\mathcal{Z}} L(f))V\xi\}({\gamma_2}^{-1}\gamma_1,\gamma_2)\\
&= \int f(\gamma_2\gamma^{-1})\{V\xi\}({\gamma_2}^{-1}\gamma_1,\gamma)d\lambda_{s(\gamma_2)}(\gamma)\\
&= \int f(\gamma_2\gamma^{-1})\xi(\gamma{\gamma_2}^{-1}\gamma_1,\gamma)d\lambda_{s(\gamma_2)}(\gamma)\\
&= \int f(\gamma_2\gamma)\xi(\gamma^{-1}{\gamma_2}^{-1}\gamma_1,\gamma^{-1})d\lambda^{s(\gamma_2)}(\gamma)\\
&= \int f(\gamma)\xi(\gamma^{-1}\gamma_1,\gamma^{-1}\gamma_2)d\lambda^{r(\gamma_2)}(\gamma).
\end{aligned}$$

Note that the system $\{L^2(\mathcal{G}^x,\lambda^x)\otimes L^2(\mathcal{G}^x,\lambda^x),\lambda(\gamma)\otimes\lambda(\gamma),\mu\}$ is a representation of the groupoid $\mathcal{G}$ in the sense of [**Ra1**], [**R**], where $\lambda(\gamma) : L^2(\mathcal{G}^x,\lambda^x) \longrightarrow L^2(\mathcal{G}^y,\lambda^y)$ $(\gamma : x \mapsto y)$

is defined to be

$$\{\lambda(\gamma)\xi\}(\gamma') = \xi(\gamma^{-1}\gamma') \qquad (\xi \in L^2(\mathcal{G}^x, \lambda^x),\ \gamma' \in \mathcal{G}).$$

It now follows from the above calculation that $\Gamma(L(f)) = V^*(1 \otimes_{\mathcal{Z}} L(f))V$ is exactly the "integrated form" of this representation of $\mathcal{G}$, while the left multiplication $L(f)$ itself is the "integrated form" of the "regular" representation $\{L^2(\mathcal{G}^x, \lambda^x), \lambda(\gamma), \mu\}$ of $\mathcal{G}$. Thus we have

$$\{\Gamma(L(f))\xi\}(\gamma_1, \gamma_2) = \int f(\gamma)\{(\lambda(\gamma) \otimes \lambda(\gamma))\xi_{s(\gamma)}\}(\gamma_1, \gamma_2) d\lambda^{r(\gamma_1)}(\gamma).$$

Suppose that $T \in \mathcal{R}(\mathcal{G})'$. Since $\mathcal{Z}_R$ is contained in $\mathcal{R}(\mathcal{G})$, T is a decomposable operator on ${}_{\mathcal{Z}}L^2(\mathcal{G}, \nu)$. Let $T = \int_X^\oplus T(x) d\mu(x)$ be the decomposition of T according to the direct integral ${}_{\mathcal{Z}}L^2(\mathcal{G}, \nu) = \int_X^\oplus L^2(\mathcal{G}^x, \lambda^x) d\mu(x)$. Then it is well-known that the family $\{T(x)\}_{x \in X}$ can be chosen so that

$$T(r(\gamma))\lambda(\gamma) = \lambda(\gamma)T(s(\gamma)), \qquad (\gamma \in \mathcal{G}).$$

Let S be another element of $\mathcal{R}(\mathcal{G})'$ and $S = \int_X^\oplus S(x) d\mu(x)$ be its decomposition. Then we have

$$S \otimes_{\mathcal{Z}} T = \int_X^\oplus S(x) \otimes T(x) d\mu(x).$$

Now, for any $f \in \mathcal{A}_I$, $\xi \in L^2(\mathcal{H}^{(2)}, \nu_1)$ and $(\gamma_1, \gamma_2) \in \mathcal{H}^{(2)}$, we obtain

$$\begin{aligned}
&\{(S \otimes_{\mathcal{Z}} T)\Gamma(L(f))\xi\}(\gamma_1, \gamma_2) \\
&= S(r(\gamma_1)) \otimes T(r(\gamma_2))\{\Gamma(L(f))\xi\}(\gamma_1, \gamma_2) \\
&= \int f(\gamma)\big(S(r(\gamma_1)) \otimes T(r(\gamma_2))\big)\{(\lambda(\gamma) \otimes \lambda(\gamma))\xi_{s(\gamma)}\}(\gamma_1, \gamma_2) d\lambda^{r(\gamma_1)}(\gamma)
\end{aligned}$$

$$= \int f(\gamma)\{(S(r(\gamma))\lambda(\gamma) \otimes T(r(\gamma))\lambda(\gamma))\xi_{s(\gamma)}\}(\gamma_1, \gamma_2) d\lambda^{r(\gamma_1)}(\gamma)$$
$$= \int f(\gamma)\{(\lambda(\gamma) \otimes \lambda(\gamma))(S(s(\gamma)) \otimes T(s(\gamma)))\xi_{s(\gamma)}\}(\gamma_1, \gamma_2) d\lambda^{r(\gamma_1)}(\gamma)$$
$$= \{\Gamma(L(f))(S \otimes_{\mathcal{Z}} T)\xi\}(\gamma_1, \gamma_2).$$

This shows that $\Gamma(L(f))$ $(f \in \mathcal{A}_I)$ commutes with the elements of the form $S \otimes_{\mathcal{Z}} T$ $(S,\, T \in \mathcal{R}(\mathcal{G})')$, which, by definition, generate a von Neumann algebra $\mathcal{R}(\mathcal{G})' \otimes_{\mathcal{Z}} \mathcal{R}(\mathcal{G})'$. Hence $\Gamma(L(f))$ belongs to the commutant $\mathcal{R}(\mathcal{G}) *_{\mathcal{Z}} \mathcal{R}(\mathcal{G})$ of $\mathcal{R}(\mathcal{G})' \otimes_{\mathcal{Z}} \mathcal{R}(\mathcal{G})'$, which is the fiber product of $\mathcal{R}(\mathcal{G})$ with itself in our terminology. Accordingly, Γ is an isomorphism from $\mathcal{R}(\mathcal{G})$ into $\mathcal{R}(\mathcal{G}) *_{\mathcal{Z}} \mathcal{R}(\mathcal{G})$.

In this paragraph, we shall show that the above defined isomorphism $\Gamma : \mathcal{R}(\mathcal{G}) \longrightarrow \mathcal{R}(\mathcal{G}) *_{\mathcal{Z}} \mathcal{R}(\mathcal{G})$ satisfies "coassociativity" in our context. So we may view this map as a "coproduct" of the algebra $\mathcal{R}(\mathcal{G})$. For this purpose, we apply Proposition 1.2 to the representations $\Gamma : \mathcal{R}(\mathcal{G}) \longrightarrow \mathcal{R}(\mathcal{G}) *_{\mathcal{Z}} \mathcal{R}(\mathcal{G})$ and $\iota : \mathcal{R}(\mathcal{G}) \longrightarrow \mathcal{R}(\mathcal{G})$ in order to obtain two homomorphisms $\Gamma *_{\mathcal{Z}} \iota$ and $\iota *_{\mathcal{Z}} \Gamma$ from $\mathcal{R}(\mathcal{G}) *_{\mathcal{Z}} \mathcal{R}(\mathcal{G})$ into $\mathcal{L}(\mathcal{H})$, where $\mathcal{H}$ is the Hilbert space ${}_{\mathcal{Z}}L^2(\mathcal{G}, \nu) \otimes_{\mu} {}_{\mathcal{Z}}L^2(\mathcal{G}, \nu) \otimes_{\mu} {}_{\mathcal{Z}}L^2(\mathcal{G}, \nu)$. What we then intend to prove is an identity:

$$(\Gamma *_{\mathcal{Z}} \iota) \circ \Gamma = (\iota *_{\mathcal{Z}} \Gamma) \circ \Gamma.$$

However, we need some preparation for this goal.

Let g be a vector in $D({}_{\mathcal{Z}}L^2(\mathcal{G}, \nu), \mu)$. We set $\|g\|_{\mu} = \|\lambda(|g|^2)\|_{\infty}$. Note that, by Lemma 2.1, $\|g\|_{\mu}$ is finite. Given such a vector g, we define an operator $T_g : L^2(\mathcal{G}, \nu) \longrightarrow {}_{\mathcal{Z}}L^2(\mathcal{G}, \nu) \otimes_{\mu} {}_{\mathcal{Z}}L^2(\mathcal{G}, \nu) \cong L^2(\mathcal{H}^{(2)}, \nu_1)$ by

$$\{T_g \xi\}(\gamma_1, \gamma_2) = g({\gamma_1}^{-1}\gamma_2)\xi(\gamma_1) \qquad (\xi \in L^2(\mathcal{G}, \nu)).$$

Then we calculate

$$\begin{aligned}
&\int\int\int |g(\gamma_1{}^{-1}\gamma_2)\xi(\gamma_1)|^2 d\lambda^{r(\gamma_1)}(\gamma_2)d\lambda^x(\gamma_1)d\mu(x) \\
&= \int\int\int |g(\gamma_2)|^2 d\lambda^{s(\gamma_1)}(\gamma_2)|\xi(\gamma_1)|^2 d\lambda^x(\gamma_1)d\mu(x) \\
&\leq \|g\|_\mu \|\xi\|^2 < \infty.
\end{aligned}$$

It follows that $T_g\xi$ indeed belongs to $L^2(\mathcal{H}^{(2)}, \nu_1)$ and that T_g is bounded with $\|T_g\| \leq \|g\|_\mu^{1/2}$. Moreover, we have

Lemma 2.5. *We have that $T_g \in \mathcal{L}_{\mathcal{R}(\mathcal{G})}(L^2(\mathcal{G}, \nu), L^2(\mathcal{H}^{(2)}, \nu_1))$. Namely, the operator T_g satisfies*

$$T_g a = \Gamma(a)T_g, \qquad (a \in \mathcal{R}(\mathcal{G})\,).$$

Furthermore, the linear span of the set $\{\, T_g\xi : g \in D({}_{\mathcal{Z}}L^2(\mathcal{G}, \nu), \mu), \ \xi \in L^2(\mathcal{G}, \nu)\,\}$ is dense in $L^2(\mathcal{H}^{(2)}, \nu_1)$.

Proof. For any $f \in \mathcal{A}_I$, we have

$$\begin{aligned}
\{T_g L(f)\xi\}(\gamma_1, \gamma_2) &= g(\gamma_1{}^{-1}\gamma_2)\{L(f)\xi\}(\gamma_1) \\
&= g(\gamma_1{}^{-1}\gamma_2)\int f(\gamma)\xi(\gamma^{-1}\gamma_1)d\lambda^{r(\gamma_1)}(\gamma).
\end{aligned}$$

We also compute

$$\begin{aligned}
\{\Gamma(L(f))T_g\xi\}(\gamma_1, \gamma_2) &= \int f(\gamma)\{T_g\xi\}(\gamma^{-1}\gamma_1, \gamma^{-1}\gamma_2)d\lambda^{r(\gamma_1)}(\gamma) \\
&= \int f(\gamma)g(\gamma_1^{-1}\gamma_2)\xi(\gamma^{-1}\gamma_1)d\lambda^{r(\gamma_1)}(\gamma).
\end{aligned}$$

This shows that $T_g L(f) = \Gamma(L(f))T_g$. for any $f \in \mathcal{A}_I$. Since the σ-weak closure of $\{\, L(f) : f \in \mathcal{A}_I \,\}$ is $\mathcal{R}(\mathcal{G})$, the first assertion follows.

For the second statement, we observe that

$$\begin{aligned}\{T_g\xi\}(\gamma_1,\gamma_2) &= \xi(\gamma_1)g(\gamma_1{}^{-1}\gamma_2)\\ &= \{W^*(\xi\otimes_\mu g)\}(\gamma_1,\gamma_2),\end{aligned}$$

where $\xi\otimes_\mu g \in L^2(\mathcal{G},\nu)_Z\otimes_{\mu Z}L^2(\mathcal{G},\nu) \cong L^2(\mathcal{G}^{(2)},\nu_2)$. Since W is a unitary and the span of $\{\,\xi\otimes_\mu g : \xi \in L^2(\mathcal{G},\nu),\ g \in D(_ZL^2(\mathcal{G},\nu),\mu)\,\}$ is dense in $L^2(\mathcal{G}^{(2)},\nu_2)$, the span of $\{\,T_g\xi : g \in D(_ZL^2(\mathcal{G},\nu),\mu),\ \xi \in L^2(\mathcal{G},\nu)\,\}$ is dense in $L^2(\mathcal{H}^{(2)},\nu_1)$. Q.E.D.

We now consider another representation $\{L^2(\mathcal{G}^x,\lambda^x)\otimes L^2(\mathcal{G}^x,\lambda^x)\otimes L^2(\mathcal{G}^x,\lambda^x), \lambda(\gamma)\otimes\lambda(\gamma)\otimes\lambda(\gamma),\mu\}$ of $\mathcal{G}$. To every $f \in \mathcal{A}_I$, we associate an "integrated form" $\hat{L}(f)$ given by the equation

$$\{\hat{L}(f)\xi\}(\gamma_1,\gamma_2,\gamma_3) = \int f(\gamma)\xi(\gamma^{-1}\gamma_1,\gamma^{-1}\gamma_2,\gamma^{-1}\gamma_3)d\lambda^{r(\gamma_1)}(\gamma),$$

where $\xi \in {}_ZL^2(\mathcal{G},\nu)\otimes_{\mu Z}L^2(\mathcal{G},\nu)\otimes_{\mu Z}L^2(\mathcal{G},\nu) \cong L^2(\mathcal{H}^{(3)},\ \tilde{\nu}_1)$. Here

$$\mathcal{H}^{(3)} = \{\,(\gamma_1,\gamma_2,\gamma_3) \in \mathcal{G}^3 : r(\gamma_1) = r(\gamma_2) = r(\gamma_3)\,\}$$

and a measure $\tilde{\nu}_1$ on $\mathcal{H}^{(3)}$ is given by the integral

$$\begin{aligned}&\int f(\gamma_1,\gamma_2,\gamma_3)d\tilde{\nu}_1(\gamma_1,\gamma_2,\gamma_3)\\ &= \int\int\int\int f(\gamma_1,\gamma_2,\gamma_3)d\lambda^x(\gamma_3)d\lambda^x(\gamma_2)d\lambda^x(\gamma_1)d\mu(x)\end{aligned}$$

for any $f \in \mathcal{F}(\mathcal{H}^{(3)})$.

Suppose that $g \in D(_ZL^2(\mathcal{G},\nu),\mu)$ and $T \in \mathcal{R}(\mathcal{G})'$. Through the direct integral $_ZL^2(\mathcal{G},\nu)=\int_X^\oplus L^2(\mathcal{G}^x,\lambda^x)d\mu(x)$, T can be decomposed into $T = \int_X^\oplus T(x)d\mu(x)$ so that we have $T(r(\gamma))\lambda(\gamma) = \lambda(\gamma)T(s(\gamma))$ $(\gamma \in \mathcal{G})$. Consider the relative tensor product $T_g\otimes_Z T$

of the above two operators. For any $f \in \mathcal{A}_I$, $\xi \in L^2(\mathcal{H}^{(2)}, \nu_1)$ and $(\gamma_1, \gamma_2, \gamma_3) \in \mathcal{H}^{(3)}$, we compute

$$\begin{aligned}
&\{(T_g \otimes_Z T)\}\Gamma(L(f))\xi\}(\gamma_1, \gamma_2, \gamma_3) \\
&= g({\gamma_1}^{-1}\gamma_2) \int f(\gamma)\{(\lambda(\gamma) \otimes T(r(\gamma))\lambda(\gamma))\xi_{s(\gamma)}\}(\gamma_1, \gamma_3) d\lambda^{r(\gamma_1)}(\gamma).
\end{aligned}$$

Also we calculate

$$\begin{aligned}
&\{\hat{L}(f)(T_g \otimes_Z T)\xi\}(\gamma_1, \gamma_2, \gamma_3) \\
&= \int f(\gamma)\{(T_g \otimes_Z T)\xi\}(\gamma^{-1}\gamma_1, \gamma^{-1}\gamma_2, \gamma^{-1}\gamma_3) d\lambda^{r(\gamma_1)}(\gamma) \\
&= \int f(\gamma) g({\gamma_1}^{-1}\gamma_2)\{(\lambda(\gamma) \otimes \lambda(\gamma) T(s(\gamma)))\xi_{s(\gamma)}\}(\gamma_1, \gamma_3) d\lambda^{r(\gamma_1)}(\gamma) \\
&= \int f(\gamma) g({\gamma_1}^{-1}\gamma_2)\{(\lambda(\gamma) \otimes T(r(\gamma))\lambda(\gamma))\xi_{s(\gamma)}\}(\gamma_1, \gamma_3) d\lambda^{r(\gamma_1)}(\gamma).
\end{aligned}$$

Thus we have shown that

$$(T_g \otimes_Z T)\Gamma(L(f))\xi = \hat{L}(f)(T_g \otimes_Z T)\xi.$$

On the other hand, in view of Proposition 1.2, we have an equality

$$(T_g \otimes_Z T)\Gamma(L(f))\xi = (\Gamma *_Z \iota)\big(\Gamma(L(f))\big)(T_g \otimes_Z T)\xi.$$

Hence it follows that $\hat{L}(f)(T_g \otimes_Z T)\xi = (\Gamma *_Z \iota)\big(\Gamma(L(f))\big)(T_g \otimes_Z T)\xi$. By Lemma 2.5, we conclude that

$$\hat{L}(f) = (\Gamma *_Z \iota)\big(\Gamma(L(f))\big) \qquad (f \in \mathcal{A}_I).$$

By a similar argument, one can also show that

$$\hat{L}(f) = (\iota *_Z \Gamma)\big(\Gamma(L(f))\big) \qquad (f \in \mathcal{A}_I).$$

Since $\{ L(f) : f \in \mathcal{A}_I \}$ is σ-weakly dense in $\mathcal{R}(\mathcal{G})$, we have proven

Theorem 2.6. *We have a "coassociative" property for Γ:*

$$(\Gamma *_{\mathcal{Z}} \iota) \circ \Gamma = (\iota *_{\mathcal{Z}} \Gamma) \circ \Gamma.$$

It is easily verified that the coproduct Γ is the usual coproduct when $\mathcal{G}$ happens to be a group.

§ 3. Actions and coactions of measured groupoids on von Neumann algebras

In this section, we will introduce notions of actions and coactions of locally compact (second countable) measured groupoids on von Neumann algebras. Before we give the definitions of these concepts, we shall briefly discuss the corresponding situation in the case of groups in order to motivate our definitions for groupoids.

Let G be a locally compact group. By an action of G on a von Neumann algebra $\mathcal{M}$, we mean a homomorphism α of G into the group $\mathrm{Aut}(\mathcal{M})$ of all automorphisms of $\mathcal{M}$ which satisfies the property that a function $s \in G \mapsto < \alpha_s(x), \omega >$ is continuous for each pair of $x \in \mathcal{M}$ and $\omega \in \mathcal{M}_*$. In the meantime, a coaction of G on a von Neumann algebra $\mathcal{N}$ is defined to be an isomorphism δ from $\mathcal{N}$ into $\mathcal{N} \bar{\otimes} \mathcal{R}(G)$ such that $\delta(1) = 1$ and $(\delta \otimes \iota) \circ \delta = (\iota \otimes \delta_G) \circ \delta$, where $\mathcal{R}(G)$ is the group von Neumann algebra of G and δ_G is an isomorphism (the so-called coproduct) of $\mathcal{R}(G)$ into $\mathcal{R}(G) \bar{\otimes} \mathcal{R}(G)$ characterized by $\delta_G(\lambda(s)) = \lambda(s) \otimes \lambda(s)$. Here λ denotes the left regular representation of G. $\mathcal{R}(G)$ can be considered as "dual object" of G from the viewpoint of the theory of Hopf-von Neumann algebras (or Kac algebras). Then a coaction of G may be regarded as an "action" of the "dual" of G. The above observation leads us to our definitions of actions and coactions of groupoids on von Neumann algebras. Before we state the definitions, we have to mention the canonical L^p-spaces associated with an arbitrary abstract von Neumann algebra. In [**Ko**], Kosaki constructed canonical L^p-spaces $L^p(\mathcal{M})$ from a given abstract von Neumann algebra (W^*-algebra) $\mathcal{M}$ without fixing any distinguished faithful normal state or any distinguished faithful normal semifintc weight on $\mathcal{M}$. In particular,

$\{\mathcal{M} = L^\infty(\mathcal{M}),\ L^2(\mathcal{M})\}$ is a standard representation. What should be noted here is that this construction is purely functorial. Thus if π is a $*$-isomorphism from $\mathcal{M}$ onto another von Neumann algebra $\mathcal{N}$, then there exists a unique unitary U from $L^2(\mathcal{M})$ onto $L^2(\mathcal{N})$ such that $\pi = \operatorname{Ad} U$. We call the unitary U the canonical implementation of π.

As before, let $(\mathcal{G},\ \{\lambda^x\},\ \Lambda,\ \delta)$ be a measured groupoid in the next definition. Remark that a groupoid $\mathcal{G}$ can be viewed not only as a generalization of groups, but also as a small category with inverses.

Definition 3.1. (1) An action of $\mathcal{G}$ is a functor $\mathcal{F}$ from $\mathcal{G}$ into the subcategory of von Neumann algebras whose arrows are surjective $*$-isomorphisms, such that, with $\mathcal{M}(x) = \mathcal{F}(x)$ $(x \in X)$ and $\alpha_\gamma = \mathcal{F}(\gamma)$ $(\gamma \in \mathcal{G})$, we impose the following two conditions on them:

(i) The family $\{\mathcal{M}(x),\ L^2(\mathcal{M}(x))\}_{x\in X}$ forms a measurable field of von Neumann algebras over $(X,\ \mu)$;

(ii) For any positive element $a = \int_X^\oplus a(x)d\mu(x)$ in the von Neumann algebra $\mathcal{M} = \int_X^\oplus \mathcal{M}(x)d\mu(x)$ and any normal positive functional $\omega = \int_X^\oplus \omega_x d\mu(x)$ in $\mathcal{M}_*^+$, the function given by $\gamma \in \mathcal{G} \longmapsto < \alpha_\gamma(a(s(\gamma))),\ \omega_{r(\gamma)} > \in \mathbf{R}_+$ is ν-measurable.

We will write $(\mathcal{G},\ \{\mathcal{M}(x)\}_{x\in X},\ \{\alpha_\gamma\}_{\gamma\in\mathcal{G}})$ for this system.

(2) Let $\{\mathcal{N}, \mathcal{K}\}$ be a von Neumann algebra. Suppose that $\mathcal{N}$ is a $\mathcal{Z}$-module. Namely, $\mathcal{Z} = L^\infty(X, \mu)$ is faithfully imbedded into $\mathcal{N}$ as a von Neumann subalgebra. Note that, in this case, one can equip the Hilbert space $\mathcal{K}$ with a $\mathcal{Z}$-module structure. Under this situation, a coaction of $\mathcal{G}$ on $\mathcal{N}$ is a $*$-isomorphism δ of $\mathcal{N}$ into the fiber product $\mathcal{N} *_{\mathcal{Z}} \mathcal{R}(\mathcal{G})$

on the Hilbert space $\mathcal{K} \otimes_{\mu \mathcal{Z}} L^2(\mathcal{G}, \nu)$ such that

$$(\delta *_{\mathcal{Z}} \iota) \circ \delta = (\iota *_{\mathcal{Z}} \Gamma) \circ \delta,$$

where Γ is the isomorphism of $\mathcal{R}(\mathcal{G})$ into $\mathcal{R}(\mathcal{G}) *_{\mathcal{Z}} \mathcal{R}(\mathcal{G})$ defined in the preceding section. We shall denote this system by $(\mathcal{G}, \mathcal{N}, \delta, \mathcal{K})$.

The author has been unable to "symmetrize" the definitions of an action and a coaction of a measured groupoid as Nakagami and Takesaki do in [**NT**]. This is because of lack of a "suitable" coproduct on $L^\infty(\mathcal{G}, \nu)$ (or this might be due to the author's ineptness of showing the existence of such a map).

We now introduce some equivalences between two actions or two coactions of a groupoid $\mathcal{G}$.

Two actions $(\mathcal{G}, \{\mathcal{M}_1(x)\}, \{\alpha_\gamma\})$ and $(\mathcal{G}, \{\mathcal{M}_2(x)\}, \{\beta_\gamma\})$ of $\mathcal{G}$ are said to be conjugate if there exists a measurable field $\{\pi_x\}_{x \in X}$ of *-isomorphisms from $\mathcal{M}_1(x)$ onto $\mathcal{M}_2(x)$ such that $\pi_{r(\gamma)} \circ \alpha_\gamma = \beta_\gamma \circ \pi_{s(\gamma)}$ for ν-a.e. $\gamma \in \mathcal{G}$. It is immediate to verify that this is an equivalence relation among actions of $\mathcal{G}$. We write $(\mathcal{G}, \{\mathcal{M}_1(x)\}, \{\alpha_\gamma\}) \cong (\mathcal{G}, \{\mathcal{M}_2(x)\}, \{\beta_\gamma\})$ if they are conjugate.

We say that two coactions $(\mathcal{G}, \mathcal{N}_1, \delta_1, \mathcal{K}_1)$ and $(\mathcal{G}, \mathcal{N}_2, \delta_2, \mathcal{K}_2)$ of $\mathcal{G}$ are conjugate, provided that there exists a $*$-isomorphism π from $\mathcal{N}_1$ onto $\mathcal{N}_2$, mapping $\mathcal{Z}$ (imbedded in $\mathcal{N}_1$) identically onto that in $\mathcal{N}_2$, such that $\delta_2 \circ \pi = (\pi *_{\mathcal{Z}} \iota) \circ \delta_1$. We write $(\mathcal{G}, \mathcal{N}_1, \delta_1, \mathcal{K}_1) \cong (\mathcal{G}, \mathcal{N}_2, \delta_2, \mathcal{K}_2)$ if this is the case. It is again an easy exercise to show that this is an equivalence relation among coactions of $\mathcal{G}$.

Let $(\mathcal{G}, \{\mathcal{M}(x)\}, \{\alpha_\gamma\})$ be an action of $\mathcal{G}$. We consider a unitary-valued function $v(\cdot)$ on $\mathcal{G}$ with the property that

(i) $v(\gamma) \in \mathcal{M}(r(\gamma))$, $v(\gamma_1\gamma_2) = v(\gamma_1)\alpha_{\gamma_1}(v(\gamma_2))$, whenever $\gamma \in \mathcal{G}$ and $(\gamma_1, \gamma_2) \in \mathcal{G}^{(2)}$.

(ii) The function $\gamma \in \mathcal{G} \mapsto < v(\gamma), \omega_{r(\gamma)} >$ is measurable for every integrable field $\{\omega_x\}_{x \in X}$ of normal functionals on $\mathcal{M}(x)$.

We set $\beta_\gamma = \mathrm{Ad}v(\gamma) \circ \alpha_\gamma$. Then one finds easily that β_γ satisfies condition (ii) of (1) in Definition 3.1. Thus we obtain a new action $(\mathcal{G}, \{\mathcal{M}(x)\}, \{\beta_\gamma\})$. We call the function v as above an α-cocycle. The action derived from $\{\beta_\gamma\}$ is called an action perturbed by an α-cocycle v. The morphism β_γ is sometimes denoted by ${}_v\alpha_\gamma$. We say that two actions are cocycle conjugate if the one is conjugate to a perturbed action of the other.

In the next two sections, we shall construct a von Neumann algebra, called the crossed product algebra, from a given action or a coaction. As one might expect, it will be proven that the above defined conjugacy (and the cocycle conjugacy in the action case) does not affect the algebraic type of this algebra.

§ 4. Crossed products by groupoid actions and their dual coactions

In the previous section, we introduced the concepts of an action and a coaction of a measured groupoid which are considered as a natural generalization of what one finds in the group case. In the case of a group action (resp. a group coaction), there is a method of obtaining a new coaction (resp. action), called the dual coaction (resp. the dual action). The method is the well-known crossed product algebra construction. We follow this idea in our groupoid setting. Thus this section is concerned with associating to a given action of a groupoid $\mathcal{G}$ a coaction of $\mathcal{G}$.

For the construction of this dual object, we start off with an action $(\mathcal{G}, \{\mathcal{M}(x)\}, \{\alpha_\gamma\})$. For each $\gamma \in \mathcal{G}$, we let $u(\gamma)$ indicate the canonical implementation of α_γ. Thus we have that $\alpha_\gamma = \mathrm{Ad}u(\gamma)$. It follows from the functoriality of canonical L^p-spaces that $u(\gamma)$ becomes a representation of $\mathcal{G}$ on the measurable field $L^2(\mathcal{M}(x))$ of Hilbert spaces over (X, μ). From now on, we write $\mathcal{H}(x)$ for the Hilbert space $L^2(\mathcal{M}(x))$ if it causes no confusion. Then, for each $x \in X$, we set $\hat{\mathcal{H}}(x) = \mathcal{H}(x) \otimes L^2(\mathcal{G}^x, \lambda^x)$. Next we define a subspace $\hat{\mathcal{M}}(\gamma)$ of $\mathcal{L}\big(\hat{\mathcal{H}}(s(\gamma)), \hat{\mathcal{H}}(r(\gamma))\big)$ by

$$\hat{\mathcal{M}}(\gamma) = \{\, au(\gamma) \otimes \lambda(\gamma) : a \in \mathcal{M}(r(\gamma)) \,\}.$$

Since $\{\hat{\mathcal{H}}(x)\}_{x \in X}$ is a measurable field of Hilbert spaces over (X, μ), we may form its direct integral:

$$\int_X^\oplus \hat{\mathcal{H}}(x) d\mu(x) = \int_X^\oplus \mathcal{H}(x) \otimes L^2(\mathcal{G}^x, \lambda^x) d\mu(x),$$

which is, by the discussion in § 1, equal to the relative tensor product $\mathcal{H} \otimes_{\mu\mathcal{Z}} L^2(\mathcal{G}, \nu)$, where $\mathcal{H} = \int_X^\oplus \mathcal{H}(x) d\mu(x)$. Note that we can identify $\mathcal{H} \otimes_{\mu\mathcal{Z}} L^2(\mathcal{G}, \nu)$ with the set of all functions

η from $\mathcal{G}$ into $\prod_{x\in X}\mathcal{H}(x)$ such that (i) $\eta(\gamma)\in\mathcal{H}(r(\gamma))$, $(\gamma\in\mathcal{G})$. (ii) the function $x\in X\mapsto\int f_{m,x}(\gamma)(\xi_{n,x}\mid\eta(\gamma))d\lambda^x(\gamma)$ is measurable for any $m,n\in\mathbf{N}$, where $\{\xi_x\}_{n\geq 1}$ and $\{f_m\}_{m\geq 1}$ are fundamental sequences of measurable fields $\{\mathcal{H}(x)\}$ and $\{L^2(\mathcal{G}^x,\lambda^x)\}$, respectively. (iii) $\int\|\eta(\gamma)\|^2 d\nu(\gamma)<\infty$. The norm of such a function η is defined by $\|\eta\|=\left(\int\|\eta(\gamma)\|^2 d\nu(\gamma)\right)^{1/2}$.

We let $\mathcal{S}(\mathcal{G},\prod_{\gamma\in\mathcal{G}}\hat{\mathcal{M}}(\gamma))$ denote the set of all sections A from $\mathcal{G}$ into $\prod_{\gamma\in\mathcal{G}}\hat{\mathcal{M}}(\gamma)$ with the following properties:

(1) If A is of the form $A(\gamma)=a(\gamma)u(\gamma)\otimes\lambda(\gamma)$ $(a(\gamma)\in\mathcal{M}(r(\gamma)))$, then the function $\gamma\in\mathcal{G}\mapsto<a(\gamma),\,\omega_{r(\gamma)}>$ is measurable for any $\omega=\int_X^{\oplus}\omega_x d\mu(x)\in\mathcal{M}_*=\int_X^{\oplus}\mathcal{M}(x)_* d\mu(x)$.

(2) The quantity $\|A\|_H=\max\{\,\|\lambda(\|A(\cdot)\|)\|_\infty,\,\|\lambda(\|A^\sharp(\cdot)\|)\|_\infty\,\}$ is finite, where $A^\sharp(\gamma)$ $=\delta(\gamma)^{-1}A(\gamma^{-1})^*$.

We will write $\mathcal{S}(\mathcal{M})=\mathcal{S}(\mathcal{G},\prod_{\gamma\in\mathcal{G}}\hat{\mathcal{M}}(\gamma))$ for short, if there is no danger of confusion. $\mathcal{S}(\mathcal{M})$ becomes a vector space under pointwise addition and scalar multiplication. We can further equip $\mathcal{S}(\mathcal{M})$ with a $\sharp$-algebra structure. To do so, we need a lemma.

Lemma 4.1.([Ha2]) *Let $I(\mathcal{G},\nu)$ be the set of all measurable functions f on $\mathcal{G}$ with $\|f\|_I<\infty$. Then we have*

$$\|f\|_I=\|f^\sharp\|_I,\qquad \|f*g\|_I\leq\|f\|_I\|g\|_I,$$

$$\|f*\xi\|_2\leq\|f\|_I\|\xi\|_2,$$

for any $f,g\in I(\mathcal{G},\nu)$ and $\xi\in L^2(\mathcal{G},\nu)$.

The proof is found in [**Ha2**].

It is an easy exercise, due to the above lemma, to show that, if one defines a product $*$ and an involution $\sharp$ by

$$(A*B)(\gamma) = \int A(\gamma_1)B(\gamma_1{}^{-1}\gamma)d\lambda^{r(\gamma)}(\gamma_1) \qquad (A, B \in \mathcal{S}(\mathcal{M})),$$

$$A^{\sharp}(\gamma) = \delta(\gamma)^{-1}A(\gamma^{-1})^*,$$

then $\mathcal{S}(\mathcal{M})$ becomes a $\sharp$-algebra.

Next we shall define a bounded representation of the algebra $\mathcal{S}(\mathcal{M})$ on the Hilbert space $\mathcal{H}\otimes_{\mu Z}L^2(\mathcal{G},\nu)$. It will be obtained by "integrating" each section in $\mathcal{S}(\mathcal{M})$.

Let A be in $\mathcal{S}(\mathcal{M})$ and $\xi, \eta \in \mathcal{H}\otimes_{\mu Z}L^2(\mathcal{G},\nu)$. We may regard ξ, η as functions on $\mathcal{G}$ as we observed before. The equation

$$(\Phi(A)\xi \mid \eta) = \int\int (a(\gamma_1)u(\gamma_1)\xi(\gamma_1{}^{-1}\gamma) \mid \eta(\gamma))d\lambda^{r(\gamma)}(\gamma_1)d\nu(\gamma)$$

defines a bounded operator $\Phi(A)$ on $\mathcal{H}\otimes_{\mu Z}L^2(\mathcal{G},\nu)$, where A has the form $A(\gamma) = a(\gamma)u(\gamma)\otimes\lambda(\gamma)$, $a(\gamma) \in \mathcal{M}(r(\gamma))$ for any $\gamma \in \mathcal{G}$. In fact, letting $f_A(\gamma) = \|A(\gamma)\| = \|a(\gamma)\|$, $f_\xi(\gamma) = \|\xi(\gamma)\|$ and $f_\eta(\gamma) = \|\eta(\gamma)\|$, we have

$$\begin{aligned}
|(\Phi(A)\xi \mid \eta)| &\leq \int\int |(a(\gamma_1)u(\gamma_1)\xi(\gamma_1{}^{-1}\gamma) \mid \eta(\gamma))|d\lambda^{r(\gamma)}(\gamma_1)d\nu(\gamma) \\
&\leq \int\int f_A(\gamma_1)f_\xi(\gamma_1{}^{-1}\gamma)f_\eta(\gamma)d\lambda^{r(\gamma)}(\gamma_1)d\nu(\gamma) \\
&= \int (f_A*f_\xi)(\gamma)f_\eta(\gamma)d\nu(\gamma).
\end{aligned}$$

Since $A \in \mathcal{S}(\mathcal{M})$, ξ and η are in $\mathcal{H}\otimes_{\mu Z}L^2(\mathcal{G},\nu)$, f_A belongs to $I(\mathcal{G},\nu)$ and $f_\xi, f_\eta \in L^2(\mathcal{G},\nu)$. From this together with the above lemma, it follows that

$$\begin{aligned}
|(\Phi(A)\xi \mid \eta)| &\leq \|f_A\|_I\|f_\xi\|_2\|f_\eta\|_2 \\
&= \|A\|_H\|\xi\|_2\|\eta\|_2.
\end{aligned}$$

Thus $\Phi(A)$ is a bounded operator on $\hat{\mathcal{H}} = \mathcal{H} \otimes_{\mu \mathcal{Z}} L^2(\mathcal{G}, \nu)$ and $\|\Phi(A)\| \leq \|A\|_H$. It can be verified further that

$$\Phi(A^\sharp) = \Phi(A)^*, \qquad \Phi(A*B) = \Phi(A)\Phi(B), \qquad (A, B \in \mathcal{S}(\mathcal{M})).$$

Moreover, we have

Lemma 4.2. *Φ is a nondegenerate norm decreasing $*$-representation of the algebra $\mathcal{S}(\mathcal{M})$ on $\hat{\mathcal{H}} = \mathcal{H} \otimes_{\mu \mathcal{Z}} L^2(\mathcal{G}, \nu)$.*

Proof. We will show the nondegeneracy of Φ. For this, we assume that $(\Phi(A)\xi \mid \eta) = 0$ for some $\eta \in \hat{\mathcal{H}}$ and any $\xi \in \hat{\mathcal{H}}$ and $A \in \mathcal{S}(\mathcal{M})$. We take particular elements of the form $\xi_n \times f$ for ξ, where $\{\xi_n\}_{n \geq 1}$ is a fundamental sequence of the measurable field $\{\mathcal{H}(x)\}_{x \in X}$ such that, if $n(x) = \dim \mathcal{H}(x)$, $\{\xi_{k,x}\}_{1 \leq k \leq n(x)}$ is an orthonormal basis of $\mathcal{H}(x)$ and $\xi_{k,x} = 0$ for $k > n(x)$, $f \in \mathcal{A}_I$, and $\xi_n \times f$ is defined to be $(\xi_n \times f)(\gamma) = f(\gamma)\xi_{n,r(\gamma)}$, $(\gamma \in \mathcal{G})$. For $A \in \mathcal{S}(\mathcal{M})$, we consider elements of the form $A_g(\gamma) = g(\gamma)u(\gamma) \otimes \lambda(\gamma)$, where $g \in \mathcal{A}_I$. Then $(\Phi(A_g)(\xi_m \times f) \mid \eta) = 0$ implies by Fubini's theorem that

$$\begin{aligned}
&\int\int f(\gamma_1^{-1}\gamma)g(\gamma_1)\big(u(\gamma_1)\xi_{m,s(\gamma_1)} \mid \eta(\gamma)\big)d\lambda^{r(\gamma)}(\gamma_1)d\nu(\gamma) \\
&= \int\int f(\gamma_1{}^{-1}\gamma)g(\gamma_1)\big(u(\gamma_1)\xi_{m,s(\gamma_1)} \mid \eta(\gamma)\big)d\lambda^{r(\gamma_1)}(\gamma)d\nu(\gamma_1) \\
&= \int\Big(\int f(\gamma_1^{-1}\gamma)\big(u(\gamma_1)\xi_{m,s(\gamma_1)} \mid \eta(\gamma)\big)d\lambda^{r(\gamma_1)}(\gamma)\Big)g(\gamma_1)d\nu(\gamma_1) \\
&= \int\Big(\int f(\gamma)\big(u(\gamma_1)\xi_{m,s(\gamma_1)}\eta(\gamma_1\gamma)\big)d\lambda_{s(\gamma_1)}(\gamma)\Big)g(\gamma_1)d\nu(\gamma_1)
\end{aligned}$$

must be zero. Since the function $\gamma_1 \in \mathcal{G} \mapsto \int f(\gamma)(u(\gamma_1)\xi_{m,s(\gamma_1)} \mid \eta(\gamma_1\gamma))d\lambda^{s(\gamma_1)}(\gamma)$ is a member of $L^2(\mathcal{G}, \nu)$ and $\mathcal{A}_I$ is dense in $L^2(\mathcal{G}, \nu)$, it follows that there exists a ν-null subset N of $\mathcal{G}$ such that

$$(*) \qquad \int f(\gamma)(u(\gamma_1)\xi_{m,s(\gamma_1)} \mid \eta(\gamma_1\gamma))d\lambda^{s(\gamma_1)}(\gamma) = 0$$

for all $\gamma_1 \in \mathcal{G} \setminus N$. Fix $\gamma_1 \in \mathcal{G} \setminus N$. Since $f \in \mathcal{A}_I$ can be arbitrary, it follows that

$$(u(\gamma_1)\xi_{m,s(\gamma_1)} \mid \eta(\gamma_1\gamma)) = 0$$

for $\lambda^{s(\gamma_1)}$-a.e. γ. Since $\{u(\gamma_1)\xi_{m,s(\gamma_1)}\}_{m\geq 1}$ is total in $\mathcal{H}(r(\gamma_1))$, it follows that $\eta(\gamma_1\gamma) = 0$ for $\lambda^{s(\gamma_1)}$-a.e. γ; so that $\int \|\eta(\gamma_1\gamma)\|^2 d\lambda^{s(\gamma_1)}(\gamma) = 0$. Let us take a positive Borel function f on $\mathcal{G}$ such that $\int f(\gamma)d\lambda^x(\gamma) = 1$ for all $x \in X$. The existence of such a function f is guaranteed by Lemma 3 in [**C3**]. Then, by Fubini's theorem, we have

$$\begin{aligned} 0 &= \int\Big(\int \|\eta(\gamma_1\gamma)\|^2 d\lambda^{s(\gamma_1)}(\gamma)\Big) f(\gamma_1)d\nu(\gamma_1) \\ &= \int\Big(\int \|\eta(\gamma)\}\|^2 d\lambda^{r(\gamma_1)}(\gamma)\Big) f(\gamma_1)d\nu(\gamma_1) \\ &= \int\int \|\eta(\gamma)\|^2 f(\gamma_1)d\lambda^{r(\gamma)}(\gamma_1)d\nu(\gamma) \\ &= \int \|\eta(\gamma)\|^2 d\nu(\gamma). \end{aligned}$$

This shows that $\eta = 0$, ν-a.e. Q.E.D.

By the above lemma, $\Phi(\mathcal{S}(\mathcal{M}))$ is a nondegenerate $*$-subalgebra of $\mathcal{L}(\hat{\mathcal{H}})$. We denote by $\mathcal{M}\times_\alpha\mathcal{G}$ the weak closure of this subalgebra. This von Neumann algebra is called the crossed product algebra obtained from the action $(\mathcal{G}, \{\mathcal{M}(x), \mathcal{H}(x)\}, \{\alpha_\gamma = \mathrm{Ad}u(\gamma)\})$. On $\hat{\mathcal{H}}$, we have a canonical $\mathcal{Z}$ action $1 \otimes_{\mathcal{Z}} M(h \circ r)$, $(h \in \mathcal{Z})$. A straightforward calculation yields an equation: $(1 \otimes_{\mathcal{Z}} M(h \circ r))\Phi(A) = \Phi(A')$, where $A'(\gamma) = h(r(\gamma))a(\gamma)u(\gamma) \otimes \lambda(\gamma)$ if $A(\gamma) = a(\gamma)u(\gamma) \otimes \lambda(\gamma)$, $(A \in \mathcal{S}(\mathcal{M}))$. Since Φ is nondegenerate, this $\mathcal{Z}$ action is contained in $\mathcal{M}\times_\alpha\mathcal{G}$. Namely, $\mathcal{Z}$ is faithfully imbedded in $\mathcal{M}\times_\alpha\mathcal{G}$ as a von Neumann subalgebra. There is another $\mathcal{Z}$ action $1 \otimes_{\mathcal{Z}} M(h \circ s)$ on $\hat{\mathcal{H}}$ which in turn commutes with the action of $\mathcal{M}\times_\alpha\mathcal{G}$.

In what follows, we shall show that there exists a coaction of $\mathcal{G}$ on the crossed product algebra $\mathcal{M}\times_\alpha\mathcal{G}$. We begin by looking at the Hilbert space $\tilde{\mathcal{H}}=\mathcal{H}\otimes_{\mu Z}L^2(\mathcal{G},\nu)\otimes_{\mu Z}L^2(\mathcal{G},\nu)$. In terms of a direct integral decomposition, we have

$$\tilde{\mathcal{H}} = \int_X^{\oplus} \mathcal{H}(x)\otimes L^2(\mathcal{G}^x,\lambda^x)\otimes L^2(\mathcal{G}^x,\lambda^x)d\mu(x).$$

Note that we may identify the Hilbert space $\tilde{\mathcal{H}}$ with the set of all functions η from $\mathcal{H}^{(2)}$ into $\prod_{x\in X}\mathcal{H}(x)$ such that (i) $\eta(\gamma_1,\gamma_2)\in\mathcal{H}(r(\gamma_1))$ for any (γ_1,γ_2) in $\mathcal{H}^{(2)}$, (ii) the function given by $x\in X\mapsto \int\int f_{k,x}(\gamma_1)f_{l,x}(\gamma_2)\big(\xi_{m,x}\mid \eta(\gamma_1,\gamma_2)\big)d\lambda^{r(\gamma_1)}(\gamma_2)d\lambda^x(\gamma_1)$ is measurable for k,l and $m\in\mathbf{N}$, where $\{f_n\}_{n\geq 1}$ and $\{\xi_m\}_{m\geq 1}$ are fundamental sequences of measurable fields for $\{L^2(\mathcal{G}^x,\lambda^x)\}_{x\in X}$ and $\{\mathcal{H}(x)\}_{x\in X}$ respectively, (iii) the function $\|\eta(\cdot)\|$ is ν_1-square-integrable: $\int\|\eta(\gamma_1,\gamma_2)\|^2d\nu_1(\gamma_1,\gamma_2)<\infty$. The norm of η is given by $\|\eta\| = \left(\int\|\eta(\gamma_1,\gamma_2)\|^2d\nu_1(\gamma_1,\gamma_2)\right)^{1/2}$. Similarly, we identify the Hilbert space $\tilde{\mathcal{H}}_1=\mathcal{H}\otimes_\mu L^2(\mathcal{G},\nu)_Z\otimes_{\mu Z}L^2(\mathcal{G},\nu)$ with the set of functions ζ from $\mathcal{G}^{(2)}$ into $\prod_{x\in X}\mathcal{H}(x)$ such that (i) $\zeta(\gamma_1,\gamma_2)\in\mathcal{H}(s(\gamma_1))$, $((\gamma_1,\gamma_2)\in\mathcal{G}^{(2)})$, (ii) the function $x\in X\mapsto \int\int \overline{Jf_{k,x}}(\gamma_1)f_{l,x}(\gamma_2)(\xi_{m,x}\mid\zeta(\gamma_1,\gamma_2))d\lambda'_x(\gamma_1)d\lambda^x(\gamma_2)$ is measurable for any k,l and $m\in\mathbf{N}$, where $\{f_n\}_{n\geq1}$ and $\{\xi_m\}_{m\geq1}$ are as above, (iii) $\int\|\zeta(\gamma_1,\gamma_2)\|^2d\nu_2(\gamma_1,\gamma_2)<\infty$. The norm of such a function ζ is $\|\zeta\| = \left(\int\|\zeta(\gamma_1,\gamma_2)\|^2d\nu_2(\gamma_1,\gamma_2)\right)^{1/2}$. Under these identifications, we state a lemma:

Lemma 4.3. *There exists a unitary $W_\mathcal{H}$ from $\tilde{\mathcal{H}}$ onto $\tilde{\mathcal{H}}_1$ defined by*

$$\{W_\mathcal{H}\eta\}(\gamma_1,\gamma_2)=u(\gamma_1)^*\eta(\gamma_1,\gamma_1\gamma_2),\qquad (\eta\in\hat{\mathcal{H}},\ (\gamma_1,\gamma_2)\in\mathcal{G}^{(2)}).$$

The inverse is given by

$$\{W_\mathcal{H}^*\zeta\}(\gamma_1,\gamma_2)=u(\gamma_1)\zeta(\gamma_1,\gamma_1{}^{-1}\gamma_2),\qquad (\zeta\in\hat{\mathcal{H}}_1,\ (\gamma_1,\gamma_2)\in\mathcal{H}^{(2)}).$$

Proof. First we compute

$$\begin{aligned}
&\int \|u(\gamma_1)^*\eta(\gamma_1, \gamma_1\gamma_2)\|^2 d\nu_2(\gamma_1, \gamma_2) \\
&= \int \|\eta(\gamma_1, \gamma_1\gamma_2)\|^2 d\lambda^{s(\gamma_1)}(\gamma_2) d\lambda^x(\gamma_1) d\mu(x) \\
&= \int \|\eta(\gamma_1, \gamma_2)\|^2 d\lambda^{r(\gamma_1)}(\gamma_2) d\lambda^x(\gamma_1) d\mu(x) \\
&= \|\eta\|^2.
\end{aligned}$$

This shows that $W_{\mathcal{H}}$ is an isometry. Also, we calculate

$$\begin{aligned}
&\int \|u(\gamma_1)\zeta(\gamma_1, {\gamma_1}^{-1}\gamma_2)\|^2 d\nu_1(\gamma_1, \gamma_2) \\
&= \int \|\zeta(\gamma_1, {\gamma_1}^{-1}\gamma_2)\|^2 d\lambda^{r(\gamma_1)}(\gamma_2) d\lambda^x(\gamma_1) d\mu(x) \\
&= \int \|\zeta(\gamma_1, \gamma_2)\|^2 d\lambda^{s(\gamma_1)}(\gamma_2) d\lambda^x(\gamma_1) d\mu(x) \\
&= \|\zeta\|^2.
\end{aligned}$$

From this, it follows that the second equation in the above assertion really defines an isometry. It is immediate to see that those isometries are inverse to each other. Hence $W_{\mathcal{H}}$ is a unitary. Q.E.D.

As we mentioned before, there is a $\mathcal{Z}$ action on $\hat{\mathcal{H}}$ given by $h \in \mathcal{Z} \mapsto 1 \otimes_{\mathcal{Z}} M(h \circ s)$. This action commutes with that of $\mathcal{M} \times_\alpha \mathcal{G}$. So we may form the relative tensor product of $\hat{\mathcal{H}}$ with this action and ${}_{\mathcal{Z}}L^2(\mathcal{G}, \nu)$. Let $\hat{\mathcal{H}}_{\mathcal{Z}} \otimes_{\mu} {}_{\mathcal{Z}}L^2(\mathcal{G}, \nu)$ denote this relative tensor product. Suppose that $\xi_1,\ \xi_2 \in \hat{\mathcal{H}}$ and $f_1,\ f_2 \in D({}_{\mathcal{Z}}L^2(\mathcal{G}, \nu), \mu)$. Then, if we let $\eta_1 = \xi_1 \otimes_\mu f_1$ and

$\eta_2 = \xi_2 \otimes_\mu f_2$, the inner product of $\hat{\mathcal{H}}_{\mathcal{Z}} \otimes_{\mu \mathcal{Z}} L^2(\mathcal{G}, \nu)$ is given by

$$
\begin{aligned}
(\eta_1 \mid \eta_2) &= (< f_2, f_2 > \xi_1 \mid \xi_2)_{\hat{\mathcal{H}}} \\
&= \int\int < f_1, f_2 > (s(\gamma))(\xi_1(\gamma) \mid \xi_2(\gamma)) d\lambda^x(\gamma) d\mu(x) \\
&= \int\int\int f_1(\gamma_1)\overline{f_2(\gamma_1)}(\xi_1(\gamma) \mid \xi_2(\gamma)) d\lambda^{s(\gamma)}(\gamma_1) d\lambda^x(\gamma) d\mu(x) \\
&= \int (f_1(\gamma_1)\xi_1(\gamma) \mid f_2(\gamma_1)\xi_2(\gamma)) d\nu_2(\gamma, \gamma_1) \\
&= \int (f_1(\gamma_1)u(\gamma)^*\xi_1(\gamma) \mid f_2(\gamma_1)u(\gamma)^*\xi_2(\gamma)) d\nu_2(\gamma, \gamma_1),
\end{aligned}
$$

This calculation shows that the equation

$$\{V_{\mathcal{H}}(\xi_1 \otimes_\mu f_1)\}(\gamma_1, \gamma_2) = f_1(\gamma_2)u(\gamma_1)^*\xi_1(\gamma_1)$$

defines an isometry from $\hat{\mathcal{H}}_{\mathcal{Z}} \otimes_{\mu \mathcal{Z}} L^2(\mathcal{G}, \nu)$ into $\tilde{\mathcal{H}}_1$. It is not hard to check that $V_{\mathcal{H}}$ is surjective. Hence $V_{\mathcal{H}}$ is a unitary. Now, since the $\mathcal{Z}$ action commutes with that of $\mathcal{M} \times_\alpha \mathcal{G}$, it makes a perfect sense to form $y \otimes_{\mathcal{Z}} 1$ on $\hat{\mathcal{H}}_{\mathcal{Z}} \otimes_{\mu \mathcal{Z}} L^2(\mathcal{G}, \nu)$ for $y \in \mathcal{M} \times_\alpha \mathcal{G}$. We define a *-isomorphism $\hat{\alpha}$ from $\mathcal{M} \times_\alpha \mathcal{G}$ into $\mathcal{L}(\tilde{\mathcal{H}})$ by

$$\hat{\alpha}(y) = W_{\mathcal{H}}^* V_{\mathcal{H}}(y \otimes_{\mathcal{Z}} 1) V_{\mathcal{H}}^* W_{\mathcal{H}}, \qquad (y \in \mathcal{M} \times_\alpha \mathcal{G}).$$

It turns out later that this morphism $\hat{\alpha}$ serves as a coaction on the crossed algebra $\mathcal{M} \times_\alpha \mathcal{G}$. For this aim, we first have to find out what $\hat{\alpha}(y)$ looks like in the special case where $y = \Phi(A)$ for some $A \in \mathcal{S}(\mathcal{M})$. To do so, for any $A \in \mathcal{S}(\mathcal{M})$, we define a section, denoted by $A \otimes \lambda$, from $\mathcal{G}$ into $\prod_{\gamma \in \mathcal{G}} \hat{\mathcal{M}}(\gamma) \otimes \lambda(\gamma)$ by $(A \otimes \lambda)(\gamma) = A(\gamma) \otimes \lambda(\gamma)$, $(\gamma \in \mathcal{G})$. To every $A \otimes \lambda$, we associate an operator $\Phi'(A \otimes \lambda)$ on $\tilde{\mathcal{H}}$ by the equation:

$$
\begin{aligned}
&(\Phi'(A \otimes \lambda)\eta_1 \mid \eta_2) \\
&= \int\int (a(\gamma)u(\gamma)\eta_1(\gamma^{-1}\gamma_1, \gamma^{-1}\gamma_2) \mid \eta_2(\gamma_1, \gamma_2)) d\lambda^{r(\gamma_1)}(\gamma) d\nu_1(\gamma_1, \gamma_2).
\end{aligned}
$$

where $A(\gamma) = a(\gamma)u(\gamma) \otimes \lambda(\gamma)$ and $\eta_i \in \tilde{\mathcal{H}}$, $(i = 1,2)$. As in the case of $\Phi(A)$, one can prove that $\|\Phi'(A \otimes \lambda)\| \leq \|A\|_H$.

Lemma 4.4. *With the above notations, we have*

$$\hat{\alpha}(\Phi(A)) = \Phi'(A \otimes \lambda), \qquad (A \in \mathcal{S}(\mathcal{M})).$$

Proof. Let $\xi \in \hat{\mathcal{H}}$, $f \in D({}_{\mathcal{Z}}L^2(\mathcal{G}, \nu), \mu)$ and $(\gamma_1, \gamma_2) \in \mathcal{H}^{(2)}$. We have

$$\begin{aligned}
&\{\Phi'(A \otimes \lambda)W_{\mathcal{H}}^* V_{\mathcal{H}}(\xi \otimes_\mu f)\}(\gamma_1, \gamma_2)\\
&= \int a(\gamma)u(\gamma)\{W_{\mathcal{H}}^* V_{\mathcal{H}}(\xi \otimes_\mu f)\}(\gamma^{-1}\gamma_1, \gamma^{-1}\gamma_2) d\lambda^{r(\gamma_1)}(\gamma)\\
&= \int a(\gamma)u(\gamma_1)\{V_{\mathcal{H}}(\xi \otimes_\mu f)\}(\gamma^{-1}\gamma_1, \gamma_1{}^{-1}\gamma_2) d\lambda^{r(\gamma_1)}(\gamma)\\
&= \int f(\gamma_1{}^{-1}\gamma_2)a(\gamma)u(\gamma_1)u(\gamma^{-1}\gamma_1)^* \xi(\gamma^{-1}\gamma_1) d\lambda^{r(\gamma_1)}(\gamma)\\
&= \int f(\gamma_1{}^{-1}\gamma_2)a(\gamma)u(\gamma)\xi(\gamma^{-1}\gamma_1) d\lambda^{r(\gamma_1)}(\gamma),
\end{aligned}$$

For $\eta \in \hat{\mathcal{H}}$, we also compute

$$\begin{aligned}
&\big(W_{\mathcal{H}}^* V_{\mathcal{H}}(\Phi(A) \otimes_{\mathcal{Z}} 1)(\xi \otimes_\mu f) \mid \eta\big)\\
&= \int \big(\{V_{\mathcal{H}}(\Phi(A)\xi \otimes_\mu f)\}(\gamma_1, \gamma_2) \mid \{W_{\mathcal{H}}\eta\}(\gamma_1, \gamma_2)\big) d\nu_2(\gamma_1, \gamma_2)\\
&= \int\int f(\gamma_1{}^{-1}\gamma_2)\big(a(\gamma)u(\gamma)\xi(\gamma^{-1}\gamma_1) \mid \eta(\gamma_1, \gamma_2)\big) d\lambda^{r(\gamma_1)}(\gamma) d\nu_1(\gamma_1, \gamma_2).
\end{aligned}$$

These calculations show that

$$\Phi'(A \otimes \lambda)W_{\mathcal{H}}^* V_{\mathcal{H}} = W_{\mathcal{H}} V_{\mathcal{H}}(\Phi(A) \otimes_{\mathcal{Z}} 1).$$

Since $\hat{\alpha}(\Phi(A)) = W_{\mathcal{H}}^* V_{\mathcal{H}}(\Phi(A) \otimes_{\mathcal{Z}} 1) V_{\mathcal{H}}^* W_{\mathcal{H}}$ by definition, the assertion of the lemma immediately follows. Q.E.D.

Next we will prove that the image of $\mathcal{M} \times_\alpha \mathcal{G}$ by the isomorphism $\hat{\alpha}$ is contained in the fiber product $(\mathcal{M} \times_\alpha \mathcal{G}) *_{\mathcal{Z}} \mathcal{R}(\mathcal{G})$ of $\mathcal{M} \times_\alpha \mathcal{G}$ and $\mathcal{R}(\mathcal{G})$ over $\mathcal{Z}$. For this purpose, we remark first that the commutant $(\mathcal{M} \times_\alpha \mathcal{G})'$ of $\mathcal{M} \times_\alpha \mathcal{G}$ on $\hat{\mathcal{H}}$ can be described explicitly as the set of decomposable operators T with decomposition $\{T(x)\}_{x\in X}$ on $\hat{\mathcal{H}} = \int_X^{\oplus} \mathcal{H}(x)\otimes L^2(\mathcal{G},\nu)d\mu(x)$ such that $T(r(\gamma))A(\gamma) = A(\gamma)T(s(\gamma))$ for ν-a.e.$\gamma \in \mathcal{G}$. Suppose now that $T \in (\mathcal{M}\times_\alpha\mathcal{G})'$ and $S \in \mathcal{R}(\mathcal{G})'$. Note that, in terms of the direct integral ${}_{\mathcal{Z}}L^2(\mathcal{G},\nu) = \int_X^{\oplus} L^2(\mathcal{G}^x,\lambda^x)d\mu(x)$, S decomposes into $S = \int_X^{\oplus} S(x)d\mu(x)$ so that we have $S(r(\gamma))\lambda(\gamma) = \lambda(\gamma)S(s(\gamma))$ for all $\gamma \in \mathcal{G}$. On $\tilde{\mathcal{H}} = \hat{\mathcal{H}}\otimes_\mu {}_{\mathcal{Z}}L^2(\mathcal{G},\nu) = \int_X^{\oplus} \mathcal{H}(x) \otimes L^2(\mathcal{G}^x,\lambda^x) \otimes L^2(\mathcal{G}^x,\lambda^x)d\mu(x)$, $T\otimes_{\mathcal{Z}}S$ is written as $T\otimes_{\mathcal{Z}}S = \int_X^{\oplus} T(x)\otimes S(x)d\mu(x)$. Let $\xi_1 = \{\xi_{1,x}\}$, $\xi_2 = \{\xi_{2,x}\} \in \hat{\mathcal{H}} = \int \mathcal{H}(x)\otimes L^2(\mathcal{G}^x,\lambda^x)d\mu(x)$ and f_1, f_2 in $D({}_{\mathcal{Z}}L^2(\mathcal{G},\nu),\mu)$. Then $\xi_1\otimes_\mu f_1$ and $\xi_2\otimes_\mu f_2$ belong to $\tilde{\mathcal{H}}$. With $A \in \mathcal{S}(\mathcal{M})$, we calculate

$$
\begin{aligned}
&\big((T\otimes_{\mathcal{Z}}S)\Phi'(A\otimes\lambda)(\xi_1\otimes_\mu f_1) \mid \xi_2\otimes_\mu f_2 \big)\\
&= \int\!\!\int \big(T(r(\gamma))A(\gamma)\xi_{1,s(\gamma)} \mid \xi_{2,r(\gamma)}\big)\\
&\qquad\qquad \big(S(r(\gamma))\lambda(\gamma)f_{1,s(\gamma)} \mid f_{2,r(\gamma)}\big)d\lambda^{r(\gamma_1)}(\gamma)d\nu_1(\gamma_1,\gamma_2)\\
&= \int\!\!\int \big(A(\gamma)T(s(\gamma))\xi_{1,s(\gamma)} \mid \xi_{2,r(\gamma)}\big)\\
&\qquad\qquad \big(\lambda(\gamma)S(s(\gamma))f_{1,s(\gamma)} \mid f_{2,r(\gamma)}\big)d\lambda(\gamma)d\nu_1(\gamma_1,\gamma_2)\\
&= \big(\Phi'(A\otimes\lambda)(T\otimes_{\mathcal{Z}}S)(\xi_1\otimes_\mu f_1) \mid \xi_2\otimes_\mu f_2\big),
\end{aligned}
$$

where $f_i = \int_X^{\oplus} f_{i,x}d\mu(x)$ in $L^2(\mathcal{G},\nu) = \int_X^{\oplus} L^2(\mathcal{G}^x,\lambda^x)d\mu(x)$, $(i = 1,2)$. It follows that $\Phi'(A\otimes\lambda)$ commutes with the elements of the form $T\otimes_{\mathcal{Z}}S$. Since such elements, by definition, generate the von Neumann algebra $(\mathcal{M}\times_\alpha\mathcal{G})'\otimes_{\mathcal{Z}}\mathcal{R}(\mathcal{G})'$, we have shown that $\hat{\alpha}(\Phi(A)) = \Phi'(A\otimes\lambda)$ belongs to $\{(\mathcal{M}\times_\alpha\mathcal{G})'\otimes_{\mathcal{Z}}\mathcal{R}(\mathcal{G})'\}' = (\mathcal{M}\times_\alpha\mathcal{G})*_{\mathcal{Z}}\mathcal{R}(\mathcal{G})$. Therefore, we conclude

Proposition 4.5. *The map $\hat{\alpha}$ defined above is a $*$-isomorphism from $\mathcal{M}\times_\alpha\mathcal{G}$ into the fiber product $(\mathcal{M}\times_\alpha\mathcal{G})*_{\mathcal{Z}}\mathcal{R}(\mathcal{G})$ over $\mathcal{Z}$.*

Our next purpose is to show that the isomorphism $\hat{\alpha}$ gives us a coaction of $\mathcal{G}$ on $\mathcal{M} \times_\alpha \mathcal{G}$. For this, we need to prove the "coassociativity" $(\hat{\alpha} *_\mathcal{Z} \iota) \circ \hat{\alpha} = (\iota *_\mathcal{Z} \Gamma) \circ \hat{\alpha}$.

By the similar method as we defined the operator $\Phi'(A \otimes \lambda)$, $(A \in \mathcal{S}(\mathcal{M}))$ on $\tilde{\mathcal{H}}$, we introduce an operator $\Phi''(A \otimes \lambda \otimes \lambda)$ on $\tilde{\mathcal{H}} \otimes_{\mu\mathcal{Z}} L^2(\mathcal{G}, \nu) = \hat{\mathcal{H}} \otimes_{\mu\mathcal{Z}} L^2(\mathcal{G}, \nu) \otimes_{\mu\mathcal{Z}} L^2(\mathcal{G}, \nu)$. So it is defined to be

$$\{\Phi''(A \otimes \lambda \otimes \lambda)(\xi \otimes_\mu f_1 \otimes_\mu f_2)\}_x$$

$$= \int A(\gamma)\xi_{s(\gamma)} \otimes \lambda(\gamma) f_{1,s(\gamma)} \otimes \lambda(\gamma) f_{2,s(\gamma)} d\lambda^x(\gamma),$$

where $\xi = \int_X^\oplus \xi_x d\mu(x)$ in $\hat{\mathcal{H}} = \int_X^\oplus \mathcal{H}(x) \otimes L^2(\mathcal{G}^x, \lambda^x) d\mu(x)$ and $f_i = \int_X^\oplus f_{i,x} d\mu(x)$ in $D({}_\mathcal{Z}L^2(\mathcal{G}, \nu), \mu)$, $(i = 1,2)$.

Lemma 4.6. $(\iota *_\mathcal{Z} \Gamma) \circ \hat{\alpha}(\Phi(A)) = \Phi''(A \otimes \lambda \otimes \lambda)$ *for any* $A \in \mathcal{S}(\mathcal{M})$.

Proof. Let $\xi_i \in \hat{\mathcal{H}}$, g_i, g and $h \in D({}_\mathcal{Z}L^2(\mathcal{G}, \nu), \mu)$, $(i = 1,2)$. Suppose that $T = \int_X^\oplus T(x) d\mu(x) \in (\mathcal{M} \times_\alpha \mathcal{G})'$. By Lemma 2.5, the operator T_g satisfies $T_g a = \Gamma(a) T_g$ $(a \in \mathcal{R}(\mathcal{G}))$. Then we compute

$$((T \otimes_\mathcal{Z} T_g)\Phi'(A \otimes \lambda)(\xi_1 \otimes_\mu g_1) \mid \xi_2 \otimes_\mu g_2 \otimes_\mu h)$$

$$= \int\int\int (\{T(r(\gamma))A(\gamma)\xi_{1,s(\gamma)}\}(\gamma_1) \mid \xi_{2,r(\gamma_1)}) g({\gamma_2}^{-1}\gamma_3) g_1(\gamma^{-1}\gamma_2)$$
$$\overline{g_2(\gamma_2)h(\gamma_3)} d\lambda^{r(\gamma_1)}(\gamma) d\lambda^{r(\gamma_2)}(\gamma_3) d\nu_1(\gamma_1, \gamma_2).$$

Also we have

$$(\Phi''(A \otimes \lambda \otimes \lambda)(T \otimes_\mathcal{Z} T_g)(\xi_1 \otimes_\mu g_1) \mid \xi_2 \otimes_\mu g_2 \otimes_\mu h)$$

$$= \int\int\int (\{A(\gamma)T(s(\gamma))\xi_{1,s(\gamma)}\}(\gamma_1) \mid \xi_{2,r(\gamma)}) \{T_g g_1\}(\gamma^{-1}\gamma_2, \gamma^{-1}\gamma_3)$$
$$\overline{g_2(\gamma_2)h(\gamma_3)} d\lambda^{r(\gamma_1)}(\gamma) d\lambda^{r(\gamma_2)}(\gamma_3) d\nu_1(\gamma_1, \gamma_2)$$

$$= \int\int\int (\{T(r(\gamma))A(\gamma)\xi_{1,s(\gamma)}\}(\gamma_1) \mid \xi_{2,r(\gamma)}) g({\gamma_2}^{-1}\gamma_3) g_1(\gamma^{-1}\gamma_2)$$
$$\overline{g_2(\gamma_2)h(\gamma_3)} d\lambda^{r(\gamma_1)}(\gamma) d\lambda^{r(\gamma_2)}(\gamma_3) d\nu_1(\gamma_1, \gamma_2).$$

Hence we conclude that

$$(T\otimes_Z T_g)\Phi'(A\otimes\lambda)(\xi_1\otimes_\mu g_1)=\Phi''(A\otimes\lambda\otimes\lambda)(T\otimes_Z T_g)(\xi_1\otimes_\mu g_1).$$

In view of Proposition 1.2, we have

$$(T\otimes_Z T_g)\Phi'(A\otimes\lambda)(\xi_1\otimes_\mu g_1)=(\iota *_Z\Gamma)(\Phi'(A\otimes\lambda))(T\otimes_Z T_g)(\xi_1\otimes_\mu g_1).$$

It follows then that

$$(\iota *_Z\Gamma)(\Phi'(A\otimes\lambda))=\Phi''(A\otimes\lambda\otimes\lambda).$$

Thus we get the desired identity. Q.E.D.

For each $g\in D({}_ZL^2(\mathcal{G},\nu),\mu)$, we define an operator R_g from $\hat{\mathcal{H}}$ into $\tilde{\mathcal{H}}$ in the same way as we defined T_g before. Namely, R_g is defined to be

$$\{R_g\eta\}(\gamma_1,\gamma_2)=g({\gamma_1}^{-1}\gamma_2)\eta(\gamma_1)\qquad(\eta\in\hat{\mathcal{H}},\ (\gamma_1,\gamma_2)\in\mathcal{H}^{(2)}).$$

Then R_g satisfies $R_g y=\hat{\alpha}(y)R_g$ for any $y\in\mathcal{M}\times_\alpha\mathcal{G}$. Also the span of $\{R_g\eta : g\in D({}_ZL^2(\mathcal{G},\nu),\mu),\ \eta\in\hat{\mathcal{H}}\}$ is dense in $\tilde{\mathcal{H}}$. The proof for this claim exactly follows the argument in Lemma 2.5. So we will not go into the details here.

Lemma 4.7. $(\hat{\alpha}*_Z\iota)\circ\hat{\alpha}(\Phi(A))=\Phi''(A\otimes\lambda\otimes\lambda)\qquad(A\in\mathcal{S}(\mathcal{M}))$.

Proof. Let $\eta_i\in\hat{\mathcal{H}}$ and $f_i,\ g_i\in D({}_ZL^2(\mathcal{G},\nu),\mu)$, $(i=1,2)$. Suppose that $S=\int_X^\oplus S(x)d\mu(x)\in\mathcal{R}(\mathcal{G})'$. Then, with $A\in\mathcal{S}(\mathcal{M})$, $A(\gamma)=a(\gamma)u(\gamma)\otimes\lambda(\gamma)$, we have

$$\begin{aligned}&\big((R_{g_1}\otimes_Z S)\Phi'(A\otimes\lambda)(\eta_1\otimes_\mu f_1)\mid\eta_2\otimes_\mu f_2\otimes_\mu g_2\big)\\&=\int\int\int g({\gamma_1}^{-1}\gamma_2)\big(a(\gamma)u(\gamma)\eta(\gamma^{-1}\gamma_1)\mid f_2(\gamma_2)\eta_2(\gamma_1)\big)\\&\qquad\{S(r(\gamma))\lambda(\gamma)f_{1,s(\gamma)}\}(\gamma_3)\overline{g_2(\gamma_3)}d\lambda^{r(\gamma)}(\gamma)d\lambda^{r(\gamma_2)}(\gamma_3)d\nu_1(\gamma_1,\gamma_2),\end{aligned}$$

Also we have

$$
\begin{aligned}
&\big(\Phi''(A\otimes\lambda\otimes\lambda)(R_{g_1}\otimes_Z S)(\eta_1\otimes_\mu f_1) \mid \eta_2\otimes_\mu f_2\otimes_\mu g_2\big)\\
&=\int\!\!\int\!\!\int (a(\gamma)u(\gamma)\{R_{g_1}\eta_1\}(\gamma^{-1}\gamma_1,\gamma^{-1}\gamma_2) \mid f_2(\gamma_2)\eta_2(\gamma_1))\\
&\qquad\{\lambda(\gamma)S(s(\gamma))f_{1,s(\gamma)}\}(\gamma_3)\overline{g_2(\gamma_3)}d\lambda^{r(\gamma_1)}(\gamma)d\lambda^{r(\gamma_2)}(\gamma_3)d\nu_1(\gamma_1,\gamma_2)\\
&=\int\!\!\int\!\!\int g(\gamma_1{}^{-1}\gamma_2)\big(a(\gamma)u(\gamma)\eta_1(\gamma^{-1}\gamma_1) \mid f_2(\gamma_2)\eta_2(\gamma_1)\big)\\
&\qquad\{S(r(\gamma))\lambda(\gamma)f_{1,s(\gamma)}\}(\gamma_3)\overline{g_2(\gamma_3)}d\lambda^{r(\gamma_1)}(\gamma)d\lambda^{r(\gamma_2)}(\gamma_3)d\nu_1(\gamma_1,\gamma_2).
\end{aligned}
$$

These calculations show that

$$
(R_{g_1}\otimes_Z S)\Phi'(A\otimes\lambda)(\eta_1\otimes_\mu f_1) = \Phi''(A\otimes\lambda\otimes\lambda)(R_{g_1}\otimes_Z S)(\eta_1\otimes_\mu f_1).
$$

On the other hand, due to Proposition 1.2, we have that $(R_{g_1}\otimes_Z S)\Phi'(A\otimes\lambda)(\eta_1\otimes_\mu f_1) = (\hat{\alpha}*_Z\iota)(\Phi'(A\otimes\lambda))(R_{g_1}\otimes_Z S)(\eta_1\otimes_\mu f_1)$. Accordingly, we obtain

$$
(\hat{\alpha}*_Z\iota)(\Phi'(A\otimes\lambda)) = \Phi''(A\otimes\lambda\otimes\lambda).
$$

Thus we are done. Q.E.D.

Thank to the combination of the last two lemmas, we may conclude the following:

Theorem 4.8. *The system $(\mathcal{G}, \mathcal{M}\times_\alpha\mathcal{G}, \hat{\alpha}, \hat{\mathcal{H}})$ is a coaction of $\mathcal{G}$ on the von Neumann algebra $\mathcal{M}\times_\alpha\mathcal{G}$. Namely, the map $\hat{\alpha}$ is a $*$-isomorphism of $\mathcal{M}\times_\alpha\mathcal{G}$ into $(\mathcal{M}\times_\alpha\mathcal{G})*_Z\mathcal{R}(\mathcal{G})$ satisfying the identity:*

$$
(\hat{\alpha}*_Z\iota)\circ\hat{\alpha} = (\iota*_Z\Gamma)\circ\hat{\alpha}.
$$

Definition 4.9. The coaction $(\mathcal{G},\ \mathcal{M}\times_\alpha \mathcal{G},\ \hat{\alpha},\ \hat{\mathcal{H}})$ obtained above is called the dual coaction of the originally given action $(\mathcal{G},\ \{\mathcal{M}(x)\},\ \{\alpha_\gamma\})$.

In the preceding paragraph, we showed that each action of $\mathcal{G}$ naturally gives rise to a coaction of $\mathcal{G}$ via the above "crossed product" construction. In what follows, we shall examine how the conjugacy among the actions of $\mathcal{G}$ affects the resulting dual object obtained through this construction. It turns out that two conjugate actions produce conjugate dual coactions. For this purpose, we shall hereafter consider systems of the form $(\mathcal{G},\ \{\mathcal{M}(x),\ \mathcal{H}(x)\}_{x\in X},\ \{\alpha_\gamma = \operatorname{Ad} u(\gamma)\}_{\gamma\in\mathcal{G}})$ in which

(1) the family $\{\mathcal{M}(x),\ \mathcal{H}(x)\}_{x\in X}$ is a measurable field of von Neumann algebras, where each $\mathcal{M}(x)$ does not necessarily act standardly on $\mathcal{H}(x)$;

(2) the pair $\{u(\gamma),\ \mathcal{H}(x)\}$ is a representation of $\mathcal{G}$ over $(X,\ \mu)$.

Let us still call such a system, including the measurable field $\{\mathcal{H}(x)\}_{x\in X}$ and the representation $u(\gamma)$ in question, an action of $\mathcal{G}$. By the same method as the one discussed so far, we can construct a coaction on a von Neumann algebra from such an action. Abusing terminologies, we shall hererafter call the new coaction a dual coaction and the new von Neumann algebra a crossed product algebra. We may also define (cocycle) conjugacy between two actions of $\mathcal{G}$ similarly.

Now let us begin by the following lemma.

Lemma 4.10. *Suppose that we are given two actions* $(\mathcal{G},\ \{\mathcal{M}(x),\mathcal{H}(x)\},\ \{\alpha_\gamma = \operatorname{Ad}u(\gamma)\})$ *and* $(\mathcal{G},\ \{\mathcal{M}(x),\mathcal{H}(x)\},\ \{\alpha_\gamma = \operatorname{Ad}v(\gamma)\})$ *of* $\mathcal{G}$*. Let* $\mathcal{N}_1$ *(resp.* $\mathcal{N}_2$ *) and* δ_1 *(resp.* δ_2 *) be the crossed product algebra and the associated dual coaction of the former (resp. the latter) action of* $\mathcal{G}$*. Then the systems* $(\mathcal{G}, \mathcal{N}_1,\ \delta_1,\ \hat{\mathcal{H}})$ *and* $(\mathcal{G}, \mathcal{N}_2,\ \delta_2,\ \hat{\mathcal{H}})$ *are* *conjugate.*

Proof. First we define a unitary U on $\hat{\mathcal{H}}$ by

$$\{U\xi\}(\gamma) = v(\gamma)u(\gamma)^*\xi(\gamma) \qquad (\xi \in \hat{\mathcal{H}},\ \gamma \in \mathcal{G}).$$

Its inverse is clearly given by $\{U^*\xi\}(\gamma) = u(\gamma)v(\gamma)^*\xi(\gamma)$. We set

$$\mathcal{S}_1(\mathcal{M}) = \mathcal{S}(\mathcal{G}, \textstyle\prod_{\gamma\in\mathcal{G}}\mathcal{M}(r(\gamma))u(\gamma)\otimes\lambda(\gamma))$$

$$\mathcal{S}_2(\mathcal{M}) = \mathcal{S}(\mathcal{G}, \textstyle\prod_{\gamma\in\mathcal{G}}\mathcal{M}(r(\gamma))v(\gamma)\otimes\lambda(\gamma)).$$

Then there is a bijection between them, sending $A_u(\gamma) = a(\gamma)u(\gamma) \otimes \lambda(\gamma)$ to $A_v(\gamma) = a(\gamma)v(\gamma) \otimes \lambda(\gamma)$. We compute

$$\begin{aligned}
&\{U\Phi(A_u)U^*\xi\}(\gamma)\\
&= v(\gamma)u(\gamma)^*\{\Phi(A_u)U^*\xi\}(\gamma)\\
&= v(\gamma)u(\gamma)^*\int a(\gamma_1)u(\gamma_1)\{U^*\xi\}({\gamma_1}^{-1}\gamma)d\lambda^{r(\gamma)}(\gamma_1)\\
&= \int v(\gamma)u(\gamma)^*a(\gamma_1)u(\gamma_1)u({\gamma_1}^{-1}\gamma)v({\gamma_1}^{-1}\gamma)^*\xi({\gamma_1}^{-1}\gamma)d\lambda^{r(\gamma)}(\gamma_1)\\
&= \int v(\gamma)u(\gamma)^*a(\gamma_1)u(\gamma)v(\gamma)^*v(\gamma_1)\xi({\gamma_1}^{-1}\gamma)d\lambda^{r(\gamma)}(\gamma_1)\\
&= \int \alpha_\gamma(\alpha_{\gamma^{-1}}(a(\gamma_1))v(\gamma_1)\xi({\gamma_1}^{-1}\gamma)d\lambda^{r(\gamma)}(\gamma_1)\\
&= \int a(\gamma_1)v(\gamma_1)\xi({\gamma_1}^{-1}\gamma)d\lambda^{r(\gamma)}(\gamma) = \{\Phi(A_v)\xi\}(\gamma).
\end{aligned}$$

Hence $U\Phi(A_u)U^* = \Phi(A_v)$, which implies that $U\mathcal{N}_1U^* = \mathcal{N}_2$. So we define a $*$-isomorphism π of $\mathcal{N}_1$ onto $\mathcal{N}_2$ by $\pi = \mathrm{Ad}U$.

Let $\xi \in \hat{\mathcal{H}}$, $f \in D({}_{\mathcal{Z}}L^2(\mathcal{G},\nu),\mu)$ and $(\gamma_1,\gamma_2) \in \mathcal{H}^{(2)}$. We take $T = \int_X^\oplus T(x)d\mu(x)$ is in $\mathcal{R}(\mathcal{G})'$. Then we have

$$\begin{aligned}
&\{(U\otimes_{\mathcal{Z}} T)\Phi'(A_u\otimes\lambda)(\xi\otimes_\mu f)\}(\gamma_1,\gamma_2)\\
&= \int\{T(r(\gamma_2))\lambda(\gamma)f_{s(\gamma)}\}(\gamma_2)v(\gamma_1)u(\gamma_1)^*a(\gamma)u(\gamma)\xi(\gamma^{-1}\gamma_1)d\lambda^{r(\gamma_1)}(\gamma)\\
&= \int\{\lambda(\gamma)T(s(\gamma))f_{s(\gamma)}\}(\gamma_2)v(\gamma_1)u(\gamma_1)^*a(\gamma)u(\gamma)\xi(\gamma^{-1}\gamma_1)d\lambda^{r(\gamma_1)}(\gamma).
\end{aligned}$$

Also we compute

$$\{\Phi'(A_v \otimes \lambda)(U \otimes_Z T)(\xi \otimes_\mu f)\}(\gamma_1, \gamma_2)$$
$$= \{\Phi'(A_v \otimes \lambda)(U\xi \otimes_\mu Tf)\}(\gamma_1, \gamma_2)$$
$$= \int \{\lambda(\gamma)T(s(\gamma))f_{s(\gamma)}\}(\gamma_2)a(\gamma)v(\gamma)\{U\xi\}(\gamma^{-1}\gamma_1)d\lambda^{r(\gamma_1)}(\gamma),$$

In the meantime, we have

$$\begin{aligned} a(\gamma)v(\gamma)\{U\xi\}(\gamma^{-1}\gamma_1) &= a(\gamma)v(\gamma)v(\gamma^{-1}\gamma_1)u(\gamma^{-1}\gamma_1)^*\xi(\gamma^{-1}\gamma_1) \\ &= v(\gamma_1)u(\gamma_1)^*u(\gamma_1)v(\gamma_1)a(\gamma)v(\gamma_1)u(\gamma_1)^*u(\gamma)\xi(\gamma^{-1}\gamma_1) \\ &= v(\gamma_1)u(\gamma_1)^*\alpha_{\gamma_1} \circ \alpha_{\gamma_1^{-1}}(a(\gamma))u(\gamma)\xi(\gamma^{-1}\gamma_1) \\ &= v(\gamma_1)u(\gamma_1)^*a(\gamma)u(\gamma)\xi(\gamma^{-1}\gamma_1). \end{aligned}$$

This shows that $(U \otimes_Z T)\Phi'(A_u \otimes \lambda)(\xi \otimes_\mu f) = \Phi'(A_v \otimes \lambda)(U \otimes_Z T)(\xi \otimes_\mu f)$. On the other hand, in view of Proposition 1.2, we have an identity: $(U \otimes_Z T)\Phi'(A_u \otimes \lambda)(\xi \otimes_\mu f) = (\pi *_Z \iota)\big(\Phi'(A_u \otimes \lambda)\big)(U \otimes_Z T)(\xi \otimes_\mu f)$. Accordingly, we have

$$(\pi *_Z \iota)\big(\Phi'(A_u \otimes \lambda)\big) = \Phi'(A_v \otimes \lambda),$$

that is,

$$(\pi *_Z \iota)(\delta_1\big(\Phi(A_u))\big) = \delta_2(\Phi(A_v)) = \delta_2 \circ \pi(\Phi(A_v)).$$

It follows that $(\pi *_Z \iota) \circ \delta_1 = \delta_2 \circ \pi$. Namely, π gives the desired conjugacy between the associated dual coactions. Q.E.D.

Let $(\mathcal{G}, \{\mathcal{M}(x), \mathcal{H}(x)\}, \{\alpha_\gamma = \mathrm{Ad}u(\gamma)\})$ be an action of $\mathcal{G}$ and $\mathcal{K}_0$ be an infinite dimensional separable Hilbert space. Then, using $\mathcal{K}_0$, we may form an action $(\mathcal{G}, \{\mathbf{C}_{\mathcal{K}_0} \otimes \mathcal{M}(x), \mathcal{K}_0 \otimes \mathcal{H}(x)\}, \{\bar{\alpha}_\gamma = \mathrm{Ad}(1 \otimes u(\gamma))\})$ of $\mathcal{G}$ conjugate to the original one.

Lemma 4.11. *The dual coactions of the above two actions are conjugate.*

Proof. Let $\mathcal{S}(\bar{\mathcal{M}}) = \mathcal{S}(\mathcal{G}, \prod_{\gamma\in\mathcal{G}}\{1_{\mathcal{K}_0}\otimes\hat{\mathcal{M}}(\gamma)\})$. Then there is a bijection between $\mathcal{S}(\mathcal{M})$ and $\mathcal{S}(\bar{\mathcal{M}})$ sending A to $1\otimes A$, where $(1\otimes A)(\gamma) = 1\otimes A(\gamma)$.

For any $\xi\in\hat{\mathcal{H}}$, $\eta\in\mathcal{K}_0$ and $A\in\mathcal{S}(\mathcal{M})$ with $A(\gamma) = a(\gamma)u(\gamma)\otimes\lambda(\gamma)$, we have

$$\begin{aligned}\{\Phi(1\otimes A)(\eta\otimes\xi)\}(\gamma) &= \int \eta\otimes a(\gamma_1)u(\gamma_1)\xi(\gamma_1{}^{-1}\gamma)d\lambda^{r(\gamma)}(\gamma_1)\\ &= \eta\otimes\int a(\gamma_1)u(\gamma_1)\xi(\gamma_1{}^{-1}\gamma)d\lambda^{r(\gamma)}(\gamma_1)\\ &= \eta\otimes\{\Phi(A)\xi\}(\gamma).\end{aligned}$$

Thus $\Phi(1\otimes A) = 1_{\mathcal{K}_0}\otimes\Phi(A)$, so that the map π given by $\pi(y) = 1\otimes y$ defines a $*$-isomorphism from $\mathcal{M}\times_\alpha\mathcal{G}$ into $\bar{\mathcal{M}}\times_\alpha\mathcal{G}$.

For every $\eta\in\mathcal{K}_0$, we may obtain a bounded operator T_η from $\hat{\mathcal{H}}$ into $\mathcal{K}_0\otimes\hat{\mathcal{H}}$ defined by $T_\eta\xi = \eta\otimes\xi$, $(\xi\in\hat{\mathcal{H}})$. It obviously satisfies the identity: $T_\eta y = \pi(y)T_\eta$ for any y in $\mathcal{M}\times_\alpha\mathcal{G}$. If $S\in\mathcal{R}(\mathcal{G})'$ with decomposition $\{S(x)\}_{x\in X}$, we have

$$\begin{aligned}&\{(T_\eta\otimes_Z S)\Phi'(A\otimes\lambda)(\xi\otimes_\mu f)\}(\gamma_1,\gamma_2)\\ &= \eta\otimes\int\{S(r(\gamma))\lambda(\gamma)f_{s(\gamma)}\}(\gamma_2)a(\gamma)u(\gamma)\xi(\gamma^{-1}\gamma_1)d\lambda^{r(\gamma_1)}(\gamma),\end{aligned}$$

whenever $\xi\in\hat{\mathcal{H}}$, $f\in D({}_ZL^2(\mathcal{G},\nu),\mu)$ and $A\in\mathcal{S}(\mathcal{M})$ with $A(\gamma) = a(\gamma)u(\gamma)\otimes\lambda(\gamma)$. Also we have

$$\begin{aligned}&\{(1\otimes\Phi'(A\otimes\lambda))(T_\eta\otimes_Z S)(\xi\otimes_\mu f)\}(\gamma_1,\gamma_2)\\ &= \{(1\otimes\Phi'(A\otimes\lambda))\big((\eta\otimes\xi)\otimes_\mu Sf\big)\}(\gamma_1,\gamma_2)\\ &= \eta\otimes\int\{\lambda(\gamma)S(s(\gamma))f_{s(\gamma)}\}(\gamma_2)a(\gamma)u(\gamma)\xi(\gamma^{-1}\gamma_1)d\lambda^{r(\gamma_1)}(\gamma)\\ &= \eta\otimes\int\{S(r(\gamma))\lambda(\gamma)f_{s(\gamma)}\}(\gamma_2)a(\gamma)u(\gamma)\xi(\gamma^{-1}\gamma_1)d\lambda^{r(\gamma_1)}(\gamma).\end{aligned}$$

This shows that

$$(T_\eta\otimes_Z S)\Phi'(A\otimes\lambda)(\xi\otimes_\mu f) = (1\otimes\Phi'(A\otimes\lambda))(T_\eta\otimes_Z S)(\xi\otimes_\mu f).$$

In the meantime, Proposition 1.2 tells us that

$$(1 \otimes \Phi'(A \otimes \lambda))(T_\eta \otimes_Z S)(\xi \otimes_\mu f) = (\pi *_Z \iota)\big(\Phi'(A \otimes \lambda)\big)(T_\eta \otimes_Z S)(\xi \otimes_\mu f).$$

It thus results that

$$(\pi *_Z \iota)\big(\hat{\alpha}(\Phi(A))\big) = 1 \otimes \Phi'(A \otimes \lambda).$$

From this, it follows that $\hat{\hat{\alpha}} \circ \pi\big(\Phi(A)\big) = (\pi *_Z \iota) \circ \hat{\alpha}\big(\Phi(A)\big)$. Therefore, the dual coactions are conjugate to each other via π. Q.E.D.

Next we consider the following situation.

Lemma 4.12. *Suppose that two actions* $(\mathcal{G}, \{\mathcal{M}_1(x), \mathcal{H}_1(x)\}, \{\alpha_\gamma = \mathrm{Ad}u(\gamma)\})$, $(\mathcal{G}, \{\mathcal{M}_2(x), \mathcal{H}_2(x)\}, \{\beta_\gamma = \mathrm{Ad}v(\gamma)\})$ *are conjugate via a measurable field* $\{\pi_x\}_{x \in X}$ *of* $*$*-isomorphisms from* $\mathcal{M}_1(x)$ *onto* $\mathcal{M}_2(x)$. *Let* $\{U_x\}_{x \in X}$ *be a measurable field of unitary operators from* $\mathcal{H}_2(x)$ *onto* $\mathcal{H}_1(x)$ *such that (i)* $U_x \pi_x(a) U_x^* = a$ *for all* $a \in \mathcal{M}_1(x)$ *and* μ*-a.e.* $x \in X$; *(ii)* $u(\gamma) U_{s(\gamma)} = U_{r(\gamma)} v(\gamma)$ *for* ν*-a.e.* $\gamma \in \mathcal{G}$. *Then the corresponding two dual coactions are conjugate.*

Proof. First note that there is a bijective correspondence between $\mathcal{S}(\mathcal{M}_1)$ and $\mathcal{S}(\mathcal{M}_2)$ via the map $A \in \mathcal{S}(\mathcal{M}_1)$ with $A(\gamma) = a(\gamma)u(\gamma) \otimes \lambda(\gamma) \mapsto \pi(A) \in \mathcal{S}(\mathcal{M}_2)$ with $\pi(A)(\gamma) = \pi_{r(\gamma)}(a(\gamma))v(\gamma) \otimes \lambda(\gamma)$.

Let $U_0 = \int_X^\oplus U_x \otimes 1 d\mu(x)$ be the unitary operator from $\int_X^\oplus \mathcal{H}_2(x) \otimes L^2(\mathcal{G}^x, \lambda^x) d\mu(x)$ onto $\int_X^\oplus \mathcal{H}_1(x) \otimes L^2(\mathcal{G}^x, \lambda^x) d\mu(x)$. In other words, U_0 is the unitary from $\hat{\mathcal{H}}_2$ onto $\hat{\mathcal{H}}_1$

given by $\{U_0\xi_2\}(\gamma) = U_{r(\gamma)}\xi_2(\gamma)$, $(\xi_2 \in \hat{\mathcal{H}}_2,\ \gamma \in \mathcal{G})$. With $\xi_2 \in \hat{\mathcal{H}}_2$ and A in $\mathcal{S}(\mathcal{M}_1)$,

$$\begin{aligned}\{U_0^*\Phi(A)U_0\xi_2\}(\gamma) &= U_{r(\gamma)}^*\{\Phi(A)U_0\xi_2\}(\gamma)\\ &= U_{r(\gamma)}^*\int a(\gamma_1)u(\gamma_1)\{U_0\xi_2\}(\gamma_1{}^{-1}\gamma)d\lambda^{r(\gamma)}(\gamma_1)\\ &= \int U_{r(\gamma)}^* a(\gamma_1)u(\gamma_1)U_{s(\gamma_1)}\xi_2(\gamma_1{}^{-1}\gamma)d\lambda^{r(\gamma)}(\gamma_1)\\ &= \int \pi_{r(\gamma_1)}(a(\gamma_1))v(\gamma_1)\xi_2(\gamma_1{}^{-1}\gamma)d\lambda^{r(\gamma)}(\gamma_1)\\ &= \{\Phi(\pi(A))\xi_2\}(\gamma),\end{aligned}$$

so $U_0^*\Phi(A)U_0 = \Phi(\pi(A))$. It follows then that $U_0^*(\mathcal{M}_1\times_\alpha\mathcal{G})U_0 = \mathcal{M}_2\times_\beta\mathcal{G}$. We set $\pi = \mathrm{Ad}U_0^*$. Note that we have $U_0^*y = \pi(y)U_0^*$ $(y \in \mathcal{M}_1\times_\alpha\mathcal{G})$.

With ξ in $\hat{\mathcal{H}}_1$, f in $D({}_ZL^2(\mathcal{G},\nu),\mu)$ and $T = \int_X^\oplus T(x)d\mu(x) \in \mathcal{R}(\mathcal{G})'$,

$$\begin{aligned}&\{(U_0^*\otimes_Z T)\Phi'(A\otimes\lambda)(\xi\otimes_\mu f)\}(\gamma_1,\gamma_2)\\ &= U_{r(\gamma)}^*\int\{T(r(\gamma))\lambda(\gamma)f_{s(\gamma)}\}(\gamma_2)a(\gamma)u(\gamma)\xi(\gamma^{-1}\gamma_1)d\lambda^{r(\gamma_1)}(\gamma)\\ &= \int\{\lambda(\gamma)T(s(\gamma))f_{s(\gamma)}\}(\gamma_2)U_{r(\gamma)}^*a(\gamma)u(\gamma)\xi(\gamma^{-1}\gamma_1)d\lambda^{r(\gamma_1)}(\gamma)\\ &= \int\{\lambda(\gamma)T(s(\gamma))f_{s(\gamma)}\}(\gamma_2)\pi_{r(\gamma)}(a(\gamma))v(\gamma)U_{s(\gamma)}^*\xi(\gamma^{-1}\gamma_1)d\lambda^{r(\gamma_1)}(\gamma)\\ &= \int\{\lambda(\gamma)T(s(\gamma))f_{s(\gamma)}\}(\gamma_2)\pi_{r(\gamma)}(a(\gamma))v(\gamma)\{U_0^*\xi\}(\gamma^{-1}\gamma_1)d\lambda^{r(\gamma_1)}(\gamma)\\ &= \{\Phi'(\pi(A)\otimes\lambda)(U_0^*\otimes_Z T)(\xi\otimes_\mu f)\}(\gamma_1,\gamma_2),\end{aligned}$$

so that the computation yields an identity: $(U_0^*\otimes_Z T)\Phi'(A\otimes\lambda)(\xi\otimes_\mu f) = \Phi'(\pi(A)\otimes\lambda)(U_0^*\otimes_Z T)(\xi\otimes_\mu f)$. Due to Proposition 1.2, we also have $(U_0^*\otimes_Z T)\Phi'(A\otimes\lambda)(\xi\otimes_\mu f) = (\pi *_Z \iota)(\Phi'(A\otimes\lambda))(U_0^*\otimes_Z T)(\xi\otimes_\mu f)$. Thus it follows that

$$(\pi *_Z \iota)\big(\Phi'(A\otimes\lambda)\big) = \Phi'(\pi(A)\otimes\lambda).$$

From this, we may conclude that

$$(\pi *_Z \iota)\circ\hat{\alpha} = \hat{\beta}\circ\pi.$$

Therefore, the corresponding dual coactions are conjugate to each other via this isomorphism π. Q.E.D.

Lemma 4.13. *Let* $(\mathcal{G},\ \{\mathcal{M}_1(x), \mathcal{H}_1(x)\},\ \{\alpha_\gamma = \mathrm{Ad}u(\gamma)\})$, $(\mathcal{G},\ \{\mathcal{M}_2(x), \mathcal{H}_2\},\ \{\beta_\gamma = \mathrm{Ad}v(\gamma)\})$ *and* $\{\pi_x\}_{x\in X}$ *be as in the previous lemma. Suppose that* $\{U_x\}_{x\in X}$ *is a family of unitary operators from* $\mathcal{H}_2(x)$ *onto* $\mathcal{H}_1(x)$ *satisfying only condition* (i) *in the previous lemma for all* $x \in X$. *Then there exists a measurable field of unitary operators with the same property as* $\{U_x\}_{x\in X}$ *for* μ*-a.e* $x \in X$.

Proof. Considering each $X_n = \{x \in X :\ \dim \mathcal{H}_1(x) = \dim \mathcal{H}_2(x) = n\}$, $n = 1,2,\ldots,\infty$, we may assume without any loss of generality that both $\{\mathcal{H}_1(x)\}$ and $\{\mathcal{H}_2(x)\}$ are a constant field $\{\mathcal{H}_1(x) = \mathcal{H}_2(x) = \mathcal{H}_0\}_{x\in X}$.

We first note that the unitary group $\mathcal{U}$ of $\mathcal{H}_0$ is complete and separable with respect to the strong* topology, hence is a Polish space.

Consider a subset B of $\mathcal{U}\times X$ defined by

$$B = \{\,(u,\, x)\ :\ \pi_x(a) = u^*au,\ \ a \in \mathcal{M}_1(x)\,\}.$$

If $\{a_n\}$ is a sequence of measurable fields of operators such that $\mathcal{M}_1(x)$ is generated by $\{\,a_n(x) : n = 1,2,\ldots\}$ for almost every $x \in X$, and if $\{\xi_n\}$ is a fundamental sequence of measurable fields of $\{\mathcal{H}_1(x)\}_{x\in X}$, then B is described by

$$B = \{\,(u,\, x) : (u\pi_x(a_n(x))\xi_{k,x} \mid \xi_{l,x}) = (a_n(x)u\xi_{k,x} \mid \xi_{l,x}) \quad \text{for any } n, k \text{ and } l \in \mathbf{N}\}.$$

Since the functions in x in the above equality are Borel after removing, if necessary, some Borel null set from X, we may regard B as a Borel subset of the standard Borel space $\mathcal{U} \times X$. By the measurable cross-section theorem (refer to [**T3**] for the details) and our assumption for $\{U_x\}_{x \in X}$, there exists a measurable function V_x on X such that $(V_x, x) \in B$, which means that

$$\pi_x(a) = V_x^* a V_x \quad (a \in \mathcal{M}_1(x)).$$

Hence this $\{V_x\}_{x \in X}$ has the desired property. Q.E.D.

Theorem 4.14. *Two conjugate actions give rise to conjugate dual coactions.*

Proof. Let $(\mathcal{G}, \{\mathcal{M}_1(x), \mathcal{H}_1(x)\}, \{\alpha_\gamma = \mathrm{Ad}u(\gamma)\})$ and $(\mathcal{G}, \{\mathcal{M}_2(x), \mathcal{H}_2(x)\}, \{\beta_\gamma = \mathrm{Ad}v(\gamma)\})$ be two actions of $\mathcal{G}$ which are conjugate via a measurable field $\{\pi_x\}_{x \in X}$ of *-isomorphisms from $\mathcal{M}_1(x)$ onto $\mathcal{M}_2(x)$.

Arguing like Proposition 3.4 of [**T2**], we may have a family $\{\mathcal{K}(x)\}_{x \in X}$ of infinite dimensional separable Hilbert spaces and a family $\{U_x\}_{x \in X}$ of unitary operators from $\mathcal{K}(x) \otimes \mathcal{H}_2(x)$ onto $\mathcal{K}(x) \otimes \mathcal{H}_1(x)$ such that $U_x(1 \otimes \pi_x(a))U_x^* = 1 \otimes a \quad (a \in \mathcal{M}_1(x),\ x \in X)$. Without any loss of generality, we may assume that all $\mathcal{K}(x)$ are equal to a fixed infinite dimensional separable Hilbert space $\mathcal{K}_0$. Next, by the previous Lemma, we may assume that $\{U_x\}_{x \in X}$ is a measurable field. Then we obtain a chain of conjugate actions as below:

$$\begin{aligned}
&(\mathcal{G}, \{\mathcal{M}_1(x), \mathcal{H}_1(x)\}, \{\alpha_\gamma = \mathrm{Ad}u(\gamma)\}) \\
&\cong (\mathcal{G}, \{\mathbf{C}_{\mathcal{K}_0} \otimes \mathcal{M}_1(x), \mathcal{K}_0 \otimes \mathcal{H}_1(x)\}, \{\mathrm{Ad}(1 \otimes u(\gamma))\}) \\
&\cong (\mathcal{G}, \{\mathbf{C}_{\mathcal{K}_0} \otimes \mathcal{M}_1(x), \mathcal{K}_0 \otimes \mathcal{H}_1(x)\}, \{\mathrm{Ad}U_{\mathrm{r}(\gamma)}(1 \otimes v(\gamma))U_{\mathrm{s}(\gamma)}^*\}) \\
&\cong (\mathcal{G}, \{\mathbf{C}_{\mathcal{K}_0} \otimes \mathcal{M}_2(x), \mathcal{K}_0 \otimes \mathcal{H}_2(x)\}, \{\mathrm{Ad}(1 \otimes v(\gamma))\}) \\
&\cong (\mathcal{G}, \{\mathcal{M}_2(x), \mathcal{H}_2(x)\}, \{\beta_\gamma = \mathrm{Ad}v(\gamma)\}).
\end{aligned}$$

Let $(\mathcal{G}, \mathcal{N}_i, \delta_i, \mathcal{K}_i)$, $(i = 1,2,\dots,5)$ be the dual coactions of respective actions. With this notation, we have

$$\begin{aligned}(\mathcal{G},\mathcal{N}_1,\delta_1,\mathcal{K}_1) &\cong (\mathcal{G},\mathcal{N}_2,\delta_2,\mathcal{K}_2) && \text{by Lemma 4.11}\\ &\cong (\mathcal{G},\mathcal{N}_3,\delta_3,\mathcal{K}_3) && \text{by Lemma 4.10}\\ &\cong (\mathcal{G},\mathcal{N}_4,\delta_4,\mathcal{K}_4) && \text{by Lemma 4.12}\\ &\cong (\mathcal{G},\mathcal{N}_5,\delta_5,\mathcal{K}_5) && \text{by Lemma 4.11.}\end{aligned}$$

Thus we are done. Q.E.D.

Theorem 4.15. *The dual coactions of two cocycle conjugate actions are conjugate.*

Proof. Let $(\mathcal{G}, \{\mathcal{M}_1(x), \mathcal{H}_1(x)\}, \{\alpha_\gamma = \mathrm{Ad}u(\gamma)\})$ and $(\mathcal{G}, \{\mathcal{M}_2(x), \mathcal{H}_2(x)\}, \{\beta_\gamma = \mathrm{Ad}u'(\gamma)\})$ be cocycle conjugate actions of $\mathcal{G}$. Thanks to the previous Theorem, we may assume that $\mathcal{M}_1(x) = \mathcal{M}_2(x)$ for all $x \in X$, and that $\beta_\gamma = {}_v\alpha_\gamma$, where v is an α-cocycle. We set $w(\gamma) = v(\gamma)u(\gamma)$. Then $\beta_\gamma = \mathrm{Ad}w(\gamma)$.

Let $\mathcal{S}_\alpha(\mathcal{M})$ (resp. $\mathcal{S}_\beta(\mathcal{M})$) be the set of sections associated with the action α (resp. β), and Φ_α (resp. Φ_β) be its representation. Suppose that $A \in \mathcal{S}_\alpha(\mathcal{M})$ with $A(\gamma) = a(\gamma)u(\gamma) \otimes \lambda(\gamma)$. For any pair of ξ, $\eta \in \hat{\mathcal{H}}$, we have

$$\begin{aligned}(\Phi_\alpha(A)\xi \mid \eta) &= \int\int (a(\gamma_1)u(\gamma_1)\xi({\gamma_1}^{-1}\gamma) \mid \eta(\gamma))d\lambda^{r(\gamma)}(\gamma_1)d\nu(\gamma)\\ &= \int\int (a(\gamma)v(\gamma_1)^*w(\gamma)\xi({\gamma_1}^{-1}\gamma) \mid \eta(\gamma))d\lambda^{r(\gamma)}(\gamma_1)d\nu(\gamma)\\ &= (\Phi_\beta(A_\beta)\xi \mid \eta),\end{aligned}$$

where $A_\beta \in \mathcal{S}_\beta(\mathcal{M})$ is defined by $A_\beta(\gamma) = a(\gamma)v(\gamma)^*w(\gamma) \otimes \lambda(\gamma)\,(= A(\gamma))$. Thus $\Phi_\alpha(A) = \Phi_\beta(A_\beta)$, which means that $\mathcal{M} \times_\alpha \mathcal{G} \subseteq \mathcal{M} \times_\beta \mathcal{G}$. Conversely, we can prove in the same way that $\mathcal{M} \times_\beta \mathcal{G} \subseteq \mathcal{M} \times_\alpha \mathcal{G}$. Hence $\mathcal{M} \times_\alpha \mathcal{G} = \mathcal{M} \times_\beta \mathcal{G}$. Q.E.D.

The above two theorems imply that the algebraic type of a crossed product algebra is independent of the choice of the representation of the groupoid $\mathcal{G}$ which implements the action $\{\alpha_\gamma\}$. The algebraic type does not depend either on how we faithfully represent the family $\{\mathcal{M}(x)\}$ on Hilbert spaces in a measurable fashion.

§ 5. Crossed products by groupoid coactions and their dual actions

In the last section, we saw that every action of $\mathcal{G}$ always gives rise to a coaction of $\mathcal{G}$ (the dual coaction), and that this association behaves nicely under the (cocycle) conjugacy. This section is in turn concerned with a construction of an action from a given coaction of $\mathcal{G}$.

We begin with a coaction $(\mathcal{G},\ \mathcal{N},\ \delta,\ \mathcal{K})$ of $\mathcal{G}$. Recall that δ is a $*$-isomorphism of $\mathcal{N}$ into the fiber product $\mathcal{N} *_{\mathcal{Z}} \mathcal{R}(\mathcal{G})$ on $\mathcal{K} \otimes_{\mu \mathcal{Z}} L^2(\mathcal{G}, \nu)$ satisfying the identity: $(\delta *_{\mathcal{Z}} \iota) \circ \delta = (\iota *_{\mathcal{Z}} \Gamma) \circ \delta$. We form a new von Neumann algebra on this Hilbert space $\mathcal{K} \otimes_{\mu \mathcal{Z}} L^2(\mathcal{G}, \nu)$, generated by $\delta(\mathcal{N})$ and $\mathbf{C} \otimes_{\mathcal{Z}} L^\infty(\mathcal{G}, \nu)$. We denote it by $\mathcal{N} \times_\delta \mathcal{G}$ and call it the crossed product algebra of $\mathcal{N}$ by the coaction δ. Note that, on $\mathcal{K} \otimes_{\mu \mathcal{Z}} L^2(\mathcal{G}, \nu)$, there is an algebra $\{\, 1 \otimes_{\mathcal{Z}} M(h \circ s) : h \in L^\infty(\mathcal{G}, \nu) \,\}$ which is a von Neumann subalgebra of $\mathbf{C} \otimes_{\mathcal{Z}} L^\infty(\mathcal{G}, \nu)$. Hence $\{\, 1 \otimes_{\mathcal{Z}} M(h \circ s) : h \in L^\infty(\mathcal{G}, \nu) \,\}$ is contained in the crossed product $\mathcal{N} \times_\delta \mathcal{G}$. Moreover, since $\mathcal{Z}_S \subseteq \mathcal{R}(\mathcal{G})'$, we have that $1 \otimes_{\mathcal{Z}} M(h \circ s) \in \mathbf{C} \otimes_{\mathcal{Z}} \mathcal{R}(\mathcal{G})' \subseteq \mathcal{N}' \otimes_{\mathcal{Z}} \mathcal{R}(\mathcal{G})' = \big(\mathcal{N} *_{\mathcal{Z}} \mathcal{R}(\mathcal{G})\big)'$. Form this together with the fact that $\delta(\mathcal{N}) \subseteq \mathcal{N} *_{\mathcal{Z}} \mathcal{R}(\mathcal{G})$, it follows that the algebra $\{\, 1 \otimes_{\mathcal{Z}} M(h \circ s) : h \in \mathcal{Z} \,\}$ is contained in the center $\mathcal{Z}(\mathcal{N} \times_\delta \mathcal{G})$ of the crossed product algebra $\mathcal{N} \times_\delta \mathcal{G}$. Thus, $\mathcal{Z} = L^\infty(X, \mu)$ can be regarded as a subalgebra of $\mathcal{Z}(\mathcal{N} \times_\delta \mathcal{G})$. We set $\mathcal{M} = \mathcal{N} \times_\delta \mathcal{G}$ and $\mathcal{H} = \mathcal{K} \otimes_{\mu \mathcal{Z}} L^2(\mathcal{G}, \nu)$. We then consider a decomposition of $\{\mathcal{M},\ \mathcal{H}\}$ relative to $\mathcal{Z}$:

$$\mathcal{M} = \int_X^\oplus \mathcal{M}(x) d\mu(x), \qquad \mathcal{H} = \int_X^\oplus \mathcal{H}(x) d\mu(x).$$

Our aim is to show that we can equip the measurable field $\{\mathcal{M}(x),\ \mathcal{H}(x)\}$ with a structure of an action of $\mathcal{G}$.

Let $\mathcal{K} = \int_X^{\oplus} \mathcal{K}(x) d\mu(x)$ be a direct integral decomposition of $\mathcal{K}$ relative to $\mathcal{Z} = L^\infty(X, \mu)$ imbedded into $\mathcal{N}$, where $\{\mathcal{K}(x)\}_{x\in X}$ can be taken to be a Borel field of Hilbert spaces. We remark that the Hilbert space $\mathcal{H} = \mathcal{K} \otimes_{\mu\mathcal{Z}} L^2(\mathcal{G}, \nu) = \int_X^{\oplus} \mathcal{K}(x) \otimes L^2(\mathcal{G}^x, \lambda^x) d\mu(x)$ can be identified with the set of all functions η from $\mathcal{G}$ into $\prod_{x\in X} \mathcal{K}(x)$ with the following properties: (i) $\eta(\gamma) \in \mathcal{K}(r(\gamma))$ $(\gamma \in \mathcal{G})$ (ii) the function $x \in X \mapsto \int f_{m,x}(\gamma)(\eta_{n,x} \mid \eta(\gamma)) d\lambda^x(\gamma)$ is μ-measurable for all m and $n \in \mathbf{N}$, where $\{\eta_n\}_{n\geq 1}$ and $\{f_m\}_{m\geq 1}$ are fundamental sequences of measurable vector fields of the families $\{\mathcal{K}(x)\}_{x\in X}$ and $\{L^2(\mathcal{G}^x, \lambda^x)\}_{x\in X}$, respectively, (iii) $\int \|\eta(\gamma)\|^2 d\nu(\gamma) < \infty$. The norm of such a function η is given by $\|\eta\| = \left(\int \|\eta(\gamma)\|^2 d\nu(\gamma)\right)^{1/2}$. Using this identification, we shall describe the above measurable field $\{\mathcal{H}(x)\}$ more explicitly. Our tool for this description is decompositions of measures. So we first need to recall a few facts about disintegrations of measures in order to make our dicussion complete. The following theorem is found in [**Ha1**], which is derived from Lemma 4.4 of [**E**].

Theorem 5.1. *Let (S, λ) be an analytic probability space, T another analytic space, and $p : S \longrightarrow T$ a Borel surjection. Suppose that a measure ν on S is equivalent to λ. Let $\tilde{\lambda} = p_*(\lambda)$ be the measure on T given by the identity: $\tilde{\lambda}(E) = \lambda(p^{-1}(E))$, where E is any Borel subset of T. Let P be a positive Borel function such that $P = d\nu/d\lambda$. Then there exist a function $t \mapsto \nu_t$ from T into the set of (σ-finite) measures on S such that*

(1) *If $f \geq 0$ is Borel on S, then $t \mapsto \int f d\nu_t$ is an extended real-valued Borel function.*

(2) *$\nu_t(S \setminus p^{-1}(\{t\})) = 0$ for all $t \in T$.*

(3) *If $f \geq 0$ is Borel on S, then $\int f d\nu = \int (\int f d\nu_t) d\tilde{\lambda}(t)$.*

The map $t \mapsto \nu_t$ *is determined by* (1), (2) *and* (3) *up to a* $\tilde{\lambda}$*-null Borel set.* $t \mapsto \nu_t$ *is determined a.e. by properties* (1) *and* (3) *together with*

(2)′ *We have* $\tilde{\lambda}(E) = \int\int 1_E \circ p d\lambda_t d\tilde{\lambda}(t)$ *for any* E *in generating subalgebra of the* σ*-field* $\mathcal{B}(T)$ *of Borel subsets in* T*. Here* 1_E *indicates the characteristic function corresponding to the subset* E*.*

Almost all λ_t *are probability measures and* $P = d\nu_t/d\lambda_t$ *a.e.*

We say that $\lambda = \int \lambda_t d\tilde{\lambda}(t)$ is a p-decompostion of λ, and that $\nu = \int \nu_t d\tilde{\lambda}(t)$ is a p-decomposition of ν with respect to $\tilde{\lambda}$. It follows that the measure class $[\nu_t]$ are determined by $[\nu]$ up to $[\tilde{\lambda}]$ null set. $[\nu]$ also determines $[\tilde{\lambda}]$.

We will make use of this Theorem to construct an action of $\mathcal{G}$ from a given coaction.

For each $x \in X$, the set $[x]$ becomes a countably generated Borel space as a subspace of the standard Borel space X. Since $\mathcal{G}_x$ is standard and the range map r is a Borel map from $\mathcal{G}_x$ onto $[x]$, $[x]$ is analytic. In particular, $[x]$ is μ-measurable. Hence $[x]$ is an analytic Borel space. On $\mathcal{G}_x$, we have a measure class $[\lambda'_x] = [\delta\lambda_x]$, so that we may consider an abelian von Neumann algebra $\mathcal{A}(x) = L^\infty(\mathcal{G}_x, [\delta\lambda_x])$. Thanks to the work [**Ko**] of Kosaki, associated to this von Neumann algebra $\mathcal{A}(x)$ is a canonical L^2-space of $\mathcal{A}(x)$, denoted by $L^2(\mathcal{G}_x, [\delta\lambda_x])$. We should remark here that this association is purely functorial. Now we have the Borel map r from $\mathcal{G}_x$ onto $[x]$. We choose a probability measure ν_x from $[\delta\lambda_x]$, and then consider the measure class $[\tilde{\nu}_x]$ on the analytic Borel space $[x]$, where $\tilde{\nu}_x = r_*(\nu_x)$. As we noted in the paragraph that follows Theorem 5.1, the measure class $[\tilde{\nu}_x]$ is independent of the choice of the measure ν_x from $[\delta\lambda_x]$. Let $\mathcal{B}(x)$ denote the abelian von Neumann algebra $L^\infty([x], [\tilde{\nu}_x])$. Since the map r is surjective,

$\mathcal{B}(x)$ is faithfully imbedded in $\mathcal{A}(x)$ as a von Neumann subalgebra. Thus $L^2(\mathcal{G}_x, [\delta\lambda_x])$ turns out to be a $\mathcal{B}(x)$-module. Let

$$L^2(\mathcal{G}_x, [\delta\lambda_x]) = \int_{[x]}^{\oplus} \mathcal{I}(x, y) d\tilde{\nu}_x(y)$$

be a direct integral decomposition of $L^2(\mathcal{G}_x, [\delta\lambda_x])$ with respect to a measure theoretic spectrum $([x], \tilde{\nu}_x)$ of $\mathcal{B}(x)$. It is always nice to have a canonical L^2-space, because it is functorially attached to our object. However, it is often convenient for the sake of proofs to work with a concrete realization of our Hilbert space $L^2(\mathcal{G}_x, [\delta\lambda_x])$. For this purpose, we will make use of Theorem 5.1. Let ν_x and $\tilde{\nu}_x$ be as above. Then we consider a r-decomposition of the measure $\delta\lambda_x$ with respect to $\tilde{\nu}_x$:

$$\delta\lambda_x = \int \tau_{x,y} d\tilde{\nu}_x(y).$$

Each measure $\tau_{x,y}$ is carried by the set $\mathcal{G}_x^y$, $(y \in [x])$. Then $L^2(\mathcal{G}_x, [\delta\lambda_x])$ is isomorphic to $L^2(\mathcal{G}_x, \delta\lambda_x)$, and the above direct integral decomposition is the same as

$$L^2(\mathcal{G}_x, \delta\lambda_x) = \int_{[x]}^{\oplus} L^2(\mathcal{G}_x^y, \tau_{x,y}) d\tilde{\nu}_x(y),$$

that is, there exists a unitary R from $L^2(\mathcal{G}_x, [\delta\lambda_x])$ onto $L^2(\mathcal{G}_x, \delta\lambda_x)$ which can be decomposed into $R = \int_{[x]}^{\oplus} R_x d\tilde{\nu}_x(y)$, where $\{R_x\}_{x \in X}$ is a measurable field of unitary operators from $\mathcal{I}(x, y)$ onto $L^2(\mathcal{G}_x^y, \tau_{x,y})$.

Let $(\mathcal{G}, \mathcal{N}, \delta, \mathcal{K})$ be a coaction of $\mathcal{G}$ as before. Let us return to the direct integral decomposition of $\mathcal{K}$ over (X, μ) again:

$$\mathcal{K} = \int_X^{\oplus} \mathcal{K}(x) d\mu(x).$$

We have the following property for the Borel field $\{\mathcal{K}(x)\}_{x\in X}$.

Lemma 5.2. *For all* $x \in X$, $\{\mathcal{K}(y)\}_{y\in[x]}$ *has the structure of a* $\tilde{\nu}_x$*-measurable field of Hilbert spaces.*

proof. Let $\{\eta_n\}_{n\geq 1}$ be a fundamental sequence of Borel vector fields of $\{\mathcal{K}(x)\}_{x\in X}$. So, for each pair of $m, n \in \mathbf{N}$, the function $f_{n,m}$ on X defined by $f_{n,m}(x) = (\eta_{n,x} \mid \eta_{m,x})$ is Borel. Thus, for each $x \in X$, $y \in [x] \mapsto (\eta_{n,y} \mid \eta_{m,y})$ is a Borel function on $[x]$. Therefore, for all $x \in X$, we can equip $\{\mathcal{K}(y)\}_{y\in[x]}$ with a $\tilde{\nu}_x$-measurable structure so that $\{\eta_{n,y}\}_{y\in[x]}$ is a fundamental sequence of vector fields of $\{\mathcal{K}(y)\}_{y\in[x]}$. Q.E.D.

Due to the above lemma, we may form a direct integral

$$\mathcal{H}'(x) = \int_{[x]}^{\oplus} L^2(\mathcal{G}_x^y, \tau_{x,y}) \otimes \mathcal{K}(y) d\tilde{\nu}_x(y).$$

We may identify the Hilbert space $\mathcal{H}'(x)$ with the set of all functions ζ from $\mathcal{G}_x$ into $\prod_{y\in[x]} \mathcal{K}(y)$ with properties: (i) $\zeta(\gamma) \in \mathcal{K}(r(\gamma))$, $(\gamma \in \mathcal{G}_x)$, (ii) the function $y \in [x] \longmapsto \int g_{m,y}(\gamma)(\eta_{n,y} \mid \zeta(\gamma)) d\tau_{x,y}(\gamma)$ is $\tilde{\nu}_x$-measurable for any $m, n \in \mathbf{N}$, where $\{g_m\}_{m\geq 1}$ is a fundamental sequence of a measurable field $\{L^2(\mathcal{G}_x^y, \tau_{x,y})\}_{y\in[x]}$ and $\{\eta_n\}_{n\geq 1}$ is as before, (iii) $\int \|\zeta(\gamma)\|^2 d\tau_{x,y}(\gamma) d\tilde{\nu}_x(y) < \infty$. The norm of ζ is defined to be $\|\zeta\| = \left(\int \|\zeta(\gamma)\|^2 d\tau_{x,y}(\gamma) d\tilde{\nu}_x(y)\right)^{1/2}$.

The family $\{\mathcal{H}'(x)\}_{x\in X}$ can be equipped with a measurable field structure in the following manner. Let $\{f_m\}_{m\geq 1}$ and $\{\eta_n\}_{n\geq 1}$ be as before. For each $x \in X$, define an element $(\tilde{f}_m \times \eta_n)_x$ in $\mathcal{H}'(x)$ by $(\tilde{f}_m \times \eta_n)_x(\gamma) = \tilde{f}_{m,x}(\gamma)\eta_{n,r(\gamma)}$, $(\gamma \in \mathcal{G}_x)$, where $\tilde{f} = J\bar{f}$. The set $\{(\tilde{f}_m \times \eta_n)_x\}_{m,n\geq 1}$ for a fixed $x \in X$ is total in $\mathcal{H}'(x)$ and, clearly, the function

$x \in X \longmapsto \big((\tilde{f}_m \times \eta_n)_x \mid (\tilde{f}_k \times \eta_l)_x \big)$ is μ-measurable for any n, m, k and l in $\mathbf{N}$. It follows that, on the family $\{\mathcal{H}'(x)\}_{x \in X}$, there exists a unique measurable field structure for which $\{(\tilde{f}_m \times \eta_n)\}_{n,m \geq 1}$ becomes a fundamental sequence of measurable vector fields. Then we can form the direct integral

$$\mathcal{H}' = \int_X^{\oplus} \mathcal{H}'(x) d\mu(x).$$

We may regard this Hilbert space $\mathcal{H}'$ as the set of functions η from $\mathcal{G}$ into $\prod_{x \in X} \mathcal{K}(x)$ satisfying the following conditions: (i) $\eta(\gamma) \in \mathcal{K}(r(\gamma))$, $(\gamma \in \mathcal{G})$, (ii) the function $x \in X \longmapsto \int \tilde{f}_{m,x}(\gamma) \ (\eta_{n,r(\gamma)} \mid \eta(\gamma)) d\lambda'_x(\gamma)$ is μ-measurable for any $m, n \in \mathbf{N}$, (iii) $\int \|\eta(\gamma)\|^2 d\nu(\gamma) < \infty$. The norm of η is defined by $\|\eta\| = \left(\int \|\eta(\gamma)\|^2 d\nu(\gamma)\right)^{1/2}$. Under this identification, the diagonal algebra $\mathcal{Z}$ acts on $\mathcal{H}'$ as follows: $\{h \cdot \eta\}(\gamma) = h(s(\gamma))\eta(\gamma)$, $(h \in \mathcal{Z}, \eta \in \mathcal{H}'$ and $\gamma \in \mathcal{G})$. Form the discussion before on the realization of elements in $\mathcal{H}$ as a certain type of functions on $\mathcal{G}$, it follows that $\mathcal{H}'$ can be viewed as the same as $\mathcal{H}$. Moreover, we may consider that the above decomposition of $\mathcal{H}'$ into a direct integral is obtained from the direct integral decomposition of $\mathcal{H}$ with respect to the algebra $\mathcal{Z}$ imbedded in the center of the crossed product algebra $\mathcal{M} = \mathcal{N} \times_\delta \mathcal{G}$. Hence, from now on, we write $\mathcal{H}(x)$, $\mathcal{H}$ for $\mathcal{H}'(x)$, $\mathcal{H}'$, respectively. We now construct a representation of $\mathcal{G}$ on the measurable field $\{\mathcal{H}(x)\}_{x \in X}$.

For every $\gamma \in \mathcal{G}$, we define a unitary $\rho_{\mathcal{K}}(\gamma)$ from $\mathcal{H}(s(\gamma))$ onto $\mathcal{H}(r(\gamma))$ by

$$\{\rho_{\mathcal{K}}(\gamma)\eta\}(\gamma_1) = \delta(\gamma)^{1/2}\eta(\gamma_1\gamma), \qquad (\eta \in \mathcal{H}(s(\gamma)), \ \gamma_1 \in \mathcal{G}_{r(\gamma)}).$$

It is an easy exercise to check that $\rho_{\mathcal{K}}$ indeed gives a representation of $\mathcal{G}$.

Let $a \in \mathcal{N}$. Then consider a decomposition:

$$\delta(a) = \int_X^{\oplus} \delta(a)_x d\mu(x)$$

of $\delta(a)$ on $\mathcal{H} = \int_X^{\oplus} \mathcal{H}(x)d\mu(x)$. Since $\delta(a)$ lies in the commutant of the von Neumann algebra arising from the representation $\rho_{\mathcal{K}}$ of $\mathcal{G}$, we may impose a condition on $\{\delta(a)_x\}_{x\in X}$: $\delta(a)_{r(\gamma)}\rho_{\mathcal{K}}(\gamma) = \rho_{\mathcal{K}}(\gamma)\delta(a)_{s(\gamma)}$, $(\gamma \in \mathcal{G})$. Namely, we have

$$\mathrm{Ad}\rho_{\mathcal{K}}(\gamma)\big(\delta(a)_{s(\gamma)}\big) = \delta(a)_{r(\gamma)}.$$

Next we consider a decomposition of $1\otimes_{\mathcal{Z}}M(f)$ $(f \in L^\infty(\mathcal{G},\nu))$ on $\mathcal{H} = \int_X^{\oplus} \mathcal{H}(x)d\mu(x)$:

$$1\otimes_{\mathcal{Z}}M(f) = \int_X^{\oplus} m_x(f_x)d\mu(x),$$

where $f_x \in L^\infty(\mathcal{G}_x, \lambda'_x)$ and $\{m_x(f_x)\xi\}(\gamma) = f_x(\gamma)\xi(\gamma)$, $(\xi \in \mathcal{H}(x),\ \gamma \in \mathcal{G}_x)$. Then, if $\gamma \in \mathcal{G}$, $f \in L^\infty(\mathcal{G}_{s(\gamma)}, \lambda'_{s(\gamma)})$ and $\xi \in \mathcal{H}(r(\gamma))$, we have

$$\begin{aligned}
\{\rho_{\mathcal{K}}(\gamma)m_{s(\gamma)}(f)\rho_{\mathcal{K}}(\gamma)^*\xi\}(\gamma_1) &= \delta(\gamma)^{1/2}\{m_{s(\gamma)}(f)\rho_{\mathcal{K}}(\gamma)^*\xi\}(\gamma_1\gamma) \\
&= \delta(\gamma)^{1/2}f(\gamma_1\gamma)\{\rho_{\mathcal{K}}(\gamma)^*\xi\}(\gamma_1\gamma) \\
&= f(\gamma_1\gamma)\xi(\gamma_1) \\
&= \{m_{r(\gamma)}\big(\rho_\gamma(f)\big)\xi\}(\gamma_1),
\end{aligned}$$

where $\rho_\gamma: L^\infty(\mathcal{G}_{s(\gamma)}, \lambda'_{s(\gamma)}) \longrightarrow L^\infty(\mathcal{G}_{r(\gamma)}, \lambda'_{r(\gamma)})$ is defined by $\rho_\gamma = \mathrm{Ad}\rho(\gamma)$ and the $\rho(\gamma)$ is the "right regular representation" of $\mathcal{G}$ giveny by

$$\{\rho(\gamma)\xi\}(\gamma_1) = \delta(\gamma)^{1/2}\xi(\gamma_1\gamma)$$

for $\xi \in L^2(\mathcal{G}_{s(\gamma)}, \lambda'_{s(\gamma)})$ and $\gamma_1 \in \mathcal{G}_{r(\gamma)}$. Hence we obtain

$$\mathrm{Ad}\rho_{\mathcal{K}}(\gamma)\big(m_{s(\gamma)}(f)\big) = m_{r(\gamma)}\big(\rho_\gamma(f)\big).$$

It follows from the above calculations that $\mathrm{Ad}\rho_{\mathcal{K}}(\gamma)$ is a $*$-isomorphism from $\mathcal{M}(s(\gamma))$ onto $\mathcal{M}(r(\gamma))$ for every $\gamma \in \mathcal{G}$, where $\{\mathcal{M}(x)\}_{x\in X}$ is, as before, a measurable field of von

Neumann algebras obtained through the decomposition of $\mathcal{M} = \mathcal{N} \times_\delta \mathcal{G}$ relative to the direct integral $\mathcal{H} = \int_X^\oplus \mathcal{H}(x)d\mu(x)$. Thus we have shown

Theorem 5.3. *The system* $(\mathcal{G},\ \{\mathcal{M}(x), \mathcal{H}(x)\},\ \{\hat{\delta}_\gamma = \mathrm{Ad}\rho_{\mathcal{K}}(\gamma)\})$ *obtained above is an action of* $\mathcal{G}$.

Definition 5.4. The above action $(\mathcal{G},\ \{\mathcal{M}(x), \mathcal{H}(x)\},\ \{\hat{\delta}_\gamma = \mathrm{Ad}\rho_{\mathcal{K}}(\gamma)\})$ is called the dual action of a given coaction $(\mathcal{G},\ \mathcal{N},\ \delta,\ \mathcal{K})$.

The definition of the dual action seems to depend upon the r-decomposition of the measure $\delta\lambda_x$ with respect to $\tilde{\nu}_x : \delta\lambda_x = \int \tau_{x,y} d\tilde{\nu}_x(y)$. However, it is easily verified by virtue of Theorem 5.1 that a different r-decomposition of $\delta\lambda_x$ gives rise to a conjugate dual action. Thus, up to conjugacy, a dual action is unique. Hence we may call it *the* dual action of $\mathcal{G}$.

We saw in the last paragraph that there is a method of associating to a coaction of $\mathcal{G}$ an action of $\mathcal{G}$ via the crossed product construction. We showed in § 4 that, in the case of actions, the crossed product construction behaves nicely in terms of (cocycle) conjugacy. In other words, two (cocycle) conjugate actions produce the two conjugate dual coactions. In what follows, we shall examine what happens in the coaction case. It turns out that the same thing is true of the case of coactions.

We start off with two coactions $(\mathcal{G},\ \mathcal{N}_1,\ \delta_1,\ \mathcal{K}_1)$ and $(\mathcal{G},\ \mathcal{N}_2,\ \delta_2,\ \mathcal{K}_2)$ conjugate via a $*$-isomorphism π from $\mathcal{N}_1$ onto $\mathcal{N}_2$. Then, arguing like Proposition 3.4 of [**T2**], we can find an infinite dimensional (separable) Hilbert space $\mathcal{K}_0$ and a unitary U from $\mathcal{K}_0 \otimes \mathcal{K}_1$

onto $\mathcal{K}_0 \otimes \mathcal{K}_2$ such that $1_{\mathcal{K}_0} \otimes \pi(a) = U(1_{\mathcal{K}_0} \otimes a)U^*$, $(a \in \mathcal{N}_1)$. We have canonical $\mathcal{Z}$ actions on both $\mathcal{K}_0 \otimes \mathcal{K}_1$ and $\mathcal{K}_0 \otimes \mathcal{K}_2$, and, under this situation, U becomes a $\mathcal{Z}$-module map. Thus, when we consider decompositions of $\mathcal{K}_0 \otimes \mathcal{K}_1$ and $\mathcal{K}_0 \otimes \mathcal{K}_2$ relative to the $\mathcal{Z}$ actions:

$$\mathcal{K}_0 \otimes \mathcal{K}_1 = \int_X^{\oplus} \mathcal{K}_0 \otimes \mathcal{K}_1(x) d\mu(x)$$

and

$$\mathcal{K}_0 \otimes \mathcal{K}_2 = \int_X^{\oplus} \mathcal{K}_0 \otimes \mathcal{K}_2(x) d\mu(x),$$

the unitary U is decomposed into a measurable field $\{U_x\}_{x \in X}$ of unitary operators from $\mathcal{K}_0 \otimes \mathcal{K}_1(x)$ onto $\mathcal{K}_0 \otimes \mathcal{K}_2(x)$. Define a unitary operator $\tilde{U}$ from $\mathcal{K}_0 \otimes \{\mathcal{K}_1 \otimes_{\mu\mathcal{Z}} L^2(\mathcal{G}, \nu)\}$ onto $\mathcal{K}_0 \otimes \{\mathcal{K}_2 \otimes_{\mu\mathcal{Z}} L^2(\mathcal{G}, \nu)\}$ by

$$\tilde{U} = \int_X^{\oplus} U_x \otimes 1_{L^2(\mathcal{G}^x, \lambda^x)} d\mu(x).$$

Let T_1 be a bounded operator from $\mathcal{K}_1$ into $\mathcal{K}_2$ with $T_1 a = \pi(a) T_1$, $(a \in \mathcal{N}_1)$. Then

$$\begin{aligned} U(1_{\mathcal{K}_0} \otimes a)U^*(1_{\mathcal{K}_0} \otimes T_1) &= (1_{\mathcal{K}_0} \otimes \pi(a))(1_{\mathcal{K}_0} \otimes T_1) \\ &= (1_{\mathcal{K}_0} \otimes T_1)(1_{\mathcal{K}_0} \otimes a). \end{aligned}$$

This shows that $U^*(1_{\mathcal{K}_0} \otimes T_1)$ commutes with $1_{\mathcal{K}_0} \otimes a$, $(a \in \mathcal{N}_1)$. Suppose that $T_2 \in \mathcal{R}(\mathcal{G})'$ and $Y \in \mathcal{N} *_{\mathcal{Z}} \mathcal{R}(\mathcal{G})$. Note that $U^*(1_{\mathcal{K}_0} \otimes T_1) \otimes_{\mathcal{Z}} T_2 = \tilde{U}^*(1_{\mathcal{K}_0} \otimes (T_1 \otimes_{\mathcal{Z}} T_2))$. Then it follows from the above calculation that

$$\begin{aligned} \tilde{U}(1_{\mathcal{K}_0} \otimes Y)\tilde{U}^*(1_{\mathcal{K}_0} \otimes (T_1 \otimes_{\mathcal{Z}} T_2)) &= \tilde{U}(1_{\mathcal{K}_0} \otimes Y)\{U^*(1_{\mathcal{K}_0} \otimes T_1) \otimes_{\mathcal{Z}} T_2\} \\ &= \tilde{U}\{U^*(1_{\mathcal{K}_0} \otimes T_1) \otimes_{\mathcal{Z}} T_2\}(1_{\mathcal{K}_0} \otimes Y) \\ &= \tilde{U}\tilde{U}^*(1_{\mathcal{K}_0} \otimes (T_1 \otimes_{\mathcal{Z}} T_2))(1_{\mathcal{K}_0} \otimes Y) \\ &= 1_{\mathcal{K}_0} \otimes (T_1 \otimes_{\mathcal{Z}} T_2) Y. \end{aligned}$$

By virtue of Proposition 1.2, we have $(\pi *_{\mathcal{Z}} \iota)(Y)(T_1 \otimes_{\mathcal{Z}} T_2) = (T_1 \otimes_{\mathcal{Z}} T_2)Y$. Thus we conclude that $\tilde{U}(1_{\mathcal{K}_0} \otimes Y)\tilde{U}^* = 1_{\mathcal{K}_0} \otimes (\pi *_{\mathcal{Z}} \iota)(Y)$ for any Y in $\mathcal{N}_1 *_{\mathcal{Z}} \mathcal{R}(\mathcal{G})$. Consequently,

$$\begin{aligned} \tilde{U}(1_{\mathcal{K}_0} \otimes \delta_1(a))\tilde{U}^* &= 1_{\mathcal{K}_0} \otimes (\pi *_{\mathcal{Z}} \iota)\big(\delta_1(a)\big) \\ &= 1_{\mathcal{K}_0} \otimes \delta_2(\pi(a)). \qquad (*) \end{aligned}$$

From the definition of $\tilde{U}$, we also have

$$\tilde{U}(1_{\mathcal{K}_0} \otimes 1_{\mathcal{K}_1} \otimes_{\mathcal{Z}} M(f))\tilde{U}^* = 1_{\mathcal{K}_0} \otimes 1_{\mathcal{K}_2} \otimes_{\mathcal{Z}} M(f)$$

for all $f \in L^\infty(\mathcal{G}, \nu)$. It is now clear that we obtain

$$\tilde{U}\big(\mathbf{C}_{\mathcal{K}_0} \otimes \mathcal{N}_1 \times_{\delta_1} \mathcal{G}\big)\tilde{U}^* = \mathbf{C}_{\mathcal{K}_0} \otimes \mathcal{N}_2 \times_{\delta_2} \mathcal{G}.$$

We define a $*$-isomorphism Π from $\mathcal{N}_1 \times_{\delta_1} \mathcal{G}$ onto $\mathcal{N}_2 \times_{\delta_2} \mathcal{G}$ by the equation

$$\tilde{U}(1 \otimes b)\tilde{U}^* = 1 \otimes \Pi(b) \qquad (b \in \mathcal{N}_1 \times_{\delta_1} \mathcal{G}).$$

Then we have

$$\Pi(\delta_1(a)) = \delta_2(\pi(a)) \qquad (a \in \mathcal{N}_1)$$

$$\Pi(1_{\mathcal{K}_1} \otimes_{\mathcal{Z}} M(f)) = 1_{\mathcal{K}_2} \otimes_{\mathcal{Z}} M(f) \qquad (f \in L^\infty(\mathcal{G}, \nu)).$$

On $\mathcal{K}_0 \otimes \{\mathcal{K}_1 \otimes_{\mu \mathcal{Z}} L^2(\mathcal{G}, \nu)\}$ (resp. on $\mathcal{K}_0 \otimes \{\mathcal{K}_0 \otimes_{\mu \mathcal{Z}} L^2(\mathcal{G}, \nu)\}$), we have a $\mathcal{Z}$ action $h \in \mathcal{Z} \longmapsto 1_{\mathcal{K}_0} \otimes (1_{\mathcal{K}_1} \otimes_{\mathcal{Z}} M(h \circ s))$ (resp. $h \in \mathcal{Z} \longmapsto 1_{\mathcal{K}_0} \otimes (1_{\mathcal{K}_2} \otimes_{\mathcal{Z}} M(h \circ s))$) which comes from the algebra $\mathcal{Z}$ imbedded in the center of $\mathcal{N}_1 \times_{\delta_1} \mathcal{G}$ (resp. $\mathcal{N}_2 \times_{\delta_2} \mathcal{G}$) via $h \in \mathcal{Z} \longmapsto 1_{\mathcal{K}_1} \otimes_{\mathcal{Z}} M(h \circ s)$ (resp. $h \in \mathcal{Z} \longmapsto 1_{\mathcal{K}_2} \otimes_{\mathcal{Z}} M(h \circ s)$) discussed at the beginning of this section. We may decompose $\mathcal{K}_0 \otimes \{\mathcal{K}_i \otimes_{\mu \mathcal{Z}} L^2(\mathcal{G}, \nu)\}$ and $\mathbf{C}_{\mathcal{K}_0} \otimes \mathcal{N}_i \times_{\delta_i} \mathcal{G}$ $(i = 1, 2)$ with respect to these $\mathcal{Z}$ actions. Then we obtain

$$\mathcal{K}_0 \otimes \{\mathcal{K}_i \otimes_{\mu \mathcal{Z}} L^2(\mathcal{G}, \nu)\} = \int_X^{\oplus} \mathcal{K}_0 \otimes \mathcal{H}_i(x)\, d\mu(x), \qquad (i = 1, 2).$$

and

$$\mathbf{C}_{\mathcal{K}_i} \otimes \mathcal{N}_i \times_{\delta_i} \mathcal{G} = \int_X^{\oplus} \mathbf{C}_{\mathcal{K}_0} \otimes \mathcal{M}_i(x) d\mu(x), \qquad (i = 1, 2).$$

Let us remark that $(\mathcal{G},\ \{\mathcal{M}_i(x),\ \mathcal{H}_i(x)\},\ \{\hat{\delta}_{i,\gamma} = \mathrm{Ad}\rho_{\mathcal{K}_i}(\gamma)\})$ are the dual actions of $(\mathcal{G},$ $\mathcal{N}_i,\ \delta_i,\ \mathcal{K}_i)$ $(i = 1, 2)$. According to these direct integrals, Π also decomposes into a measurable field $\{\pi_x\}_{x \in X}$ of $*$-isomorphisms from $\mathcal{M}_1(x)$ onto $\mathcal{M}_2(x)$. For any $a \in \mathcal{N}_1$, let $\delta_1(a) = \int_X^{\oplus} \delta_1(a)_x d\mu(x)$ (resp. $\delta_2(\pi(a)) = \int_X^{\oplus} \delta_2(\pi(a))_x d\mu(x)$) be a direct integral decomposition of $\delta_1(a)$ (resp. $\delta_2(\pi(a))$). Also, for any $f \in L^\infty(\mathcal{G}, \nu)$, let

$$1_{\mathcal{K}_i} \otimes_{\mathcal{Z}} M(f) = \int_X^{\oplus} m_x^{(i)}(f_x) d\mu(x)$$

be, as before, a direct integral decomposition of $1_{\mathcal{K}_i} \otimes_{\mathcal{Z}} M(f)$ relative to $\mathcal{K}_i \otimes_{\mu \mathcal{Z}} L^2(\mathcal{G}, \nu)$ $= \int_X^{\oplus} \mathcal{H}_i(x) d\mu(x)$ $(i = 1, 2)$. (See the notation introduced in the discussion preceding Theorem 5.3). We choose sequences $\{a_n\}_{n \geq 1}$ and $\{f_n\}_{n \geq 1}$ of elements in $\mathcal{N}_1$ and $L^\infty(\mathcal{G}, \nu)$ that generate $\mathcal{N}_1$ and $L^\infty(\mathcal{G}, \nu)$, respectively. Since

$$\begin{aligned} \int_X^{\oplus} \pi_x(\delta_1(a)_x) d\mu(x) &= \Pi(\delta_1(a)) \\ &= \delta_2(\pi(a)) \\ &= \int_X^{\oplus} \delta_2(\pi(a))_x d\mu(x), \end{aligned}$$

it follows that, for any $n \geq 1$, there exists a μ-null subset $N_1(n)$ of X such that

$$\pi_x(\delta_1(a_n)_x) = \delta_2(\pi(a_n))_x$$

for all $x \in X \setminus N_1(n)$. It also follows that, for any $n \geq 1$, there exists a μ-null subset $N_2(n)$ of X such that

$$\pi_x(m_x^{(1)}(f_{n,x})) = m_x^{(2)}(f_{n,x})$$

for all $x \in X \setminus N_2(n)$. Put $N = \cup_{n,m\geq 1} N_1(n) \cup N_2(m)$, which is a μ-null subset of X. If $\gamma \in \mathcal{G}_{X\setminus N}$, the reduction of $\mathcal{G}$ to $X \setminus N$, then we have

$$\begin{aligned}
\pi_{r(\gamma)} \circ \hat{\delta}_{1,\gamma}(\delta_1(a_n)_{s(\gamma)}) &= \pi_{r(\gamma)}(\delta_1(a_n)_{r(\gamma)}) \\
&= \delta_2(\pi(a_n))_{r(\gamma)} \\
&= \hat{\delta}_{2,\gamma}(\delta_2(\pi(a_n))_{s(\gamma)}) \\
&= \hat{\delta}_{2,\gamma} \circ \pi_{s(\gamma)}(\delta_1(a_n)_{s(\gamma)}),
\end{aligned}$$

for any $n \geq 1$. Moreover, if γ is as above, then we obtain

$$\begin{aligned}
\pi_{r(\gamma)} \circ \hat{\delta}_{1,\gamma}(m^{(1)}_{s(\gamma)}(f_{m,s(\gamma)})) &= \pi_{r(\gamma)}(m^{(1)}_{r(\gamma)}(\rho_\gamma(f_{m,s(\gamma)}))) \\
&= m^{(2)}_{r(\gamma)}(\rho_\gamma(f_{m,s(\gamma)})) \\
&= \hat{\delta}_{2,\gamma}(m^{(2)}_{s(\gamma)}(f_{m,s(\gamma)})) \\
&= \hat{\delta}_{2,\gamma} \circ \pi_{s(\gamma)}(m^{(1)}_{s(\gamma)}(f_{m,s(\gamma)})),
\end{aligned}$$

for any $m \geq 1$. Accordingly, we conclude that $\pi_{r(\gamma)} \circ \hat{\delta}_{1,\gamma} = \hat{\delta}_{2,\gamma} \circ \pi_{s(\gamma)}$ for any $\gamma \in \mathcal{G}_{X\setminus N}$. Since $\mathcal{G}_{X\setminus N}$ is ν-conull, we have proven

Theorem 5.5. *Two conjugate coactions of $\mathcal{G}$ produce conjugate dual actions.*

This theorem tells us that the algebraic type of the crossed product algebras depends only on the isomorphism class of the von Neumann algebra $\mathcal{N}$. In other words, the algebraic type is independent of how we faithfully represent the von Neumann algebra $\mathcal{N}$ on a Hilbert space.

§ 6. Duality for actions on von Neumann algebras

In the last two sections, we saw that, to a given action (resp. coaction), we can always associate a coaction (resp. an action) of $\mathcal{G}$, called the dual coaction (resp. the dual action) of $\mathcal{G}$. Then a question naturally arises whether a Takesaki duality type theorem holds in our groupoid setting. This section is devoted to answering this question affirmatively for actions of groupoids. As a corollary, we will obtain a Nakagami-Takesaki duality for actions of locally compact groups on von Neumann algebras.

We start with an action $(\mathcal{G},\ \{\mathcal{M}(x), \mathcal{H}(x)\},\ \{\alpha_\gamma = \mathrm{Ad}u(\gamma)\})$ of $\mathcal{G}$. Then the associated dual coaction is $(\mathcal{G},\ \mathcal{M} \times_\alpha \mathcal{G},\ \hat{\alpha},\ \mathcal{H} \otimes_{\mu \mathcal{Z}} L^2(\mathcal{G}, \nu))$. The crossed product algebra obtained from the dual coaction $\hat{\alpha}$ is by definition the von Neumann algebra $\{\mathcal{M} \times_\alpha \mathcal{G}\} \times_{\hat{\alpha}} \mathcal{G}$ generated by the sets $\hat{\alpha}(\mathcal{M} \times_\alpha \mathcal{G})$ and $\mathbf{C} \otimes_{\mathcal{Z}} L^\infty(\mathcal{G}, \nu)$ acting on the Hilbert space $\mathcal{H} \otimes_{\mu \mathcal{Z}} L^2(\mathcal{G}, \nu) \otimes_{\mu \mathcal{Z}} L^2(\mathcal{G}, \nu)$. Our first aim is to show that this algebra is isomorphic to the algebra $\mathcal{M} \otimes_{\mathcal{Z}} \mathcal{Z}_S'$ on the Hilbert space $\mathcal{H} \otimes_\mu L^2(\mathcal{G}, \nu)_{\mathcal{Z}}$, where $\mathcal{H} = \int_X^\oplus \mathcal{H}(x) d\mu(x)$ and $\mathcal{M} = \int_X^\oplus \mathcal{M}(x) d\mu(x)$.

Let $\tilde{\mathcal{H}}_2 = \mathcal{H} \otimes_\mu L^2(\mathcal{G}, \nu)_{\mathcal{Z}} \otimes_\mu L^2(\mathcal{G}, \nu)_{\mathcal{Z}}$. Then, as we did before, we may identify this Hilbert space $\tilde{\mathcal{H}}_2$ with the set of functions η from $\mathcal{I}^{(2)}$ into $\prod_{x \in X} \mathcal{H}(x)$ with properties that (i) $\eta(\gamma_1, \gamma_2) \in \mathcal{H}(s(\gamma_1))$ for any $(\gamma_1, \gamma_2) \in \mathcal{I}^{(2)}$; (ii) the function $x \in X \longmapsto \int\int \tilde{f}_{k,x}(\gamma_1)\, \tilde{f}_{l,x}(\gamma_2)(\xi_{m,x} \mid \eta(\gamma_1, \gamma_2)) d\lambda_x'(\gamma_1) d\lambda_x'(\gamma_2)$ is μ-measurable for any k, l and $m \in \mathbf{N}$, where $\{f_n\}_{n \geq 1}$ and $\{\xi_m\}_{m \geq 1}$ are, as before, fundamental sequences of vector fields of $\{L^2(\mathcal{G}^x, \lambda^x)\}_{x \in X}$ and $\{\mathcal{H}(x)\}_{x \in X}$, respectively; (iii) $\int \|\eta(\gamma_1, \gamma_2)\|^2 d\nu_4(\gamma_1, \gamma_2) < \infty$. The norm of such a function η is given by $\|\eta\| = \left(\int \|\eta(\gamma_1, \gamma_2)\|^2 d\nu_4(\gamma_1, \gamma_2)\right)^{1/2}$. With this identification, we may define a unitary $U_{\mathcal{H}}$ from $\tilde{\mathcal{H}}_2$ onto $\tilde{\mathcal{H}}_1$ by the equation:

$$\{U_{\mathcal{H}}\eta\}(\gamma_1, \gamma_2) = \delta(\gamma_2)^{1/2}u(\gamma_2)\eta(\gamma_1\gamma_2, \gamma_2) \qquad ((\gamma_1, \gamma_2) \in \mathcal{G}^{(2)}).$$

Exactly the same way as we proved in § 2 that U is unitary, one can verify that the above operator $U_{\mathcal{H}}$ is indeed a unitary. Its inverse is given by

$$\{U_{\mathcal{H}}^{*}\xi\}(\gamma_1, \gamma_2) = \delta(\gamma_2)^{-1/2}u(\gamma_2)^{*}\xi(\gamma_1{\gamma_2}^{-1}, \gamma_2) \qquad ((\gamma_1, \gamma_2) \in \mathcal{I}^{(2)}).$$

As we noted before, there is a $\mathcal{Z}$ action on $\hat{\mathcal{H}}$ given by $h \in \mathcal{Z} \longmapsto 1 \otimes_{\mathcal{Z}} M(h \circ s)$. Recall that, in § 4, we wrote $\hat{\mathcal{H}}_{\mathcal{Z}}$ for the Hilbert space $\hat{\mathcal{H}}$ when we consider this $\mathcal{Z}$ action on it particularly. Suppose that ξ_1, $\xi_2 \in \hat{\mathcal{H}}$ and g_1, $g_2 \in D(L^2(\mathcal{G}, \nu)_{\mathcal{Z}}, \mu)$. Then the inner product of $\hat{\mathcal{H}}_{\mathcal{Z}} \otimes_{\mu} L^2(\mathcal{G}, \nu)_{\mathcal{Z}}$ is defined to be

$$\begin{aligned}
&(\xi_1 \otimes_{\mu} g_1 \mid \xi_2 \otimes_{\mu} g_2) \\
&= (< g_1, g_2 >^{\circ} \xi_1 \mid \xi_2) \\
&= \int < g_1, g_2 >^{\circ} (s(\gamma))(\xi_1(\gamma) \mid \xi_2(\gamma)) d\nu(\gamma) \\
&= \iiint g_1(\gamma_1)\overline{g_2(\gamma_1)}(\xi_1(\gamma) \mid \xi_2(\gamma)) d\lambda'_{s(\gamma)}(\gamma_1) d\lambda^{x}(\gamma) d\mu(x) \\
&= \iiint (g_1(\gamma_1)u(\gamma)^{*}\xi_1(\gamma) \mid g_2(\gamma_1)u(\gamma)^{*}\xi_2(\gamma)) d\lambda'_{x}(\gamma_1) d\lambda'_{x}(\gamma) d\mu(x) \\
&= \int (g_1(\gamma_1)u(\gamma)^{*}\xi_1(\gamma) \mid g_2(\gamma_1)u(\gamma)^{*}\xi_2(\gamma)) d\nu_4(\gamma, \gamma_1).
\end{aligned}$$

(Refer to § 2 for the definition of the notation $< g_1, g_2 >^{\circ}$.) The above computation implies that the equation

$$\{K_{\mathcal{H}}(\xi_1 \otimes_{\mu} g_1)\}(\gamma_1, \gamma_2) = g_1(\gamma_2)u(\gamma)^{*}\xi_1(\gamma_1) \qquad ((\gamma_1, \gamma_2) \in \mathcal{I}^{(2)})$$

defines an isometry from $\hat{\mathcal{H}}_{\mathcal{Z}} \otimes_{\mu} L^2(\mathcal{G}, \nu)_{\mathcal{Z}}$ into $\tilde{\mathcal{H}}_2$. One can also check that $K_{\mathcal{H}}$ is surjective. Hence $K_{\mathcal{H}}$ is a unitary.

Next we look at the Hilbert space $\mathcal{H}\otimes_\mu L^2(\mathcal{G},\nu)_{\mathcal{Z}}$. In terms of a direct integral decomposition, we have

$$\mathcal{H}\otimes_\mu L^2(\mathcal{G},\nu)_{\mathcal{Z}} = \int_X^{\oplus} \mathcal{H}(x)\otimes L^2(\mathcal{G}_{\mathrm{x}},\lambda'_{\mathrm{x}})\,d\mu(x).$$

We regard this space as the set of functions η from $\mathcal{G}$ into $\prod_{x\in X}\mathcal{H}(x)$ with (i) $\eta(\gamma)\in\mathcal{H}(s(\gamma))$, $(\gamma\in\mathcal{G})$; (ii) the function $x\in X\longmapsto\int\tilde{f}_{m,x}(\gamma)(\xi_{n,x}\mid\eta(\gamma))d\lambda'_x(\gamma)$ is μ-measurable for any m, $n\in\mathbf{N}$, where $\{\xi_n\}$ and $\{f_m\}$ are as before; (iii) $\int\|\eta(\gamma)\|^2d\nu(\gamma)<\infty$. The quantity $\left(\int\|\eta(\gamma)\|^2d\nu(\gamma)\right)^{1/2}$ gives the norm of η. We define a unitary P from $\mathcal{H}\otimes_\mu L^2(\mathcal{G},\nu)_{\mathcal{Z}}$ onto $\hat{\mathcal{H}}$ by

$$\{P\eta\}(\gamma)=u(\gamma)\eta(\gamma)\qquad(\eta\in\mathcal{H}\otimes_\mu L^2(\mathcal{G},\nu)_{\mathcal{Z}},\ \gamma\in\mathcal{G}).$$

We consider a $\mathcal{Z}$ action on $\mathcal{H}\otimes_\mu L^2(\mathcal{G},\nu)_{\mathcal{Z}}$ given by $h\in\mathcal{Z}\longmapsto 1\otimes_{\mathcal{Z}}M(h\circ s)$. We denote by ρ this representation of $\mathcal{Z}$, that is, $\rho(h)=1\otimes_{\mathcal{Z}}M(h\circ s)$. Then the unitary P is a $\mathcal{Z}$-module map from $\mathcal{H}\otimes_\mu L^2(\mathcal{G},\nu)_{\mathcal{Z}}$ with the above action ρ onto $\hat{\mathcal{H}}_{\mathcal{Z}}$: indeed, for $\eta\in\mathcal{H}\otimes_\mu L^2(\mathcal{G},\nu)_{\mathcal{Z}}$, $h\in\mathcal{Z}$ and $\gamma\in\mathcal{G}$, we have

$$\begin{aligned}\{(1\otimes_{\mathcal{Z}}M(h\circ s))P\eta\}(\gamma)&=h(s(\gamma))\{P\eta\}(\gamma)\\&=h(s(\gamma))u(\gamma)\eta(\gamma)\\&=\{P\rho(h)\eta\}(\gamma),\end{aligned}$$

which implies that $(1\otimes_{\mathcal{Z}}M(h\circ s))P=P\rho(h)$.

Let y be in $\mathcal{M}\otimes_{\mathcal{Z}}\mathcal{Z}'_{\mathcal{S}}$ on $\mathcal{H}\otimes_\mu L^2(\mathcal{G},\nu)_{\mathcal{Z}}$. Then it follows from the preceding paragraph that PyP^* commutes with the $\mathcal{Z}$ action on $\hat{\mathcal{H}}_{\mathcal{Z}}$. Thus it makes a good sense to form $PyP^*\otimes_{\mathcal{Z}}1$ on $\hat{\mathcal{H}}_{\mathcal{Z}}\otimes_\mu L^2(\mathcal{G},\nu)_{\mathcal{Z}}$. We now define a $*$-isomorphism π from $\mathcal{M}\otimes_{\mathcal{Z}}\mathcal{Z}'_{\mathcal{S}}$ into $\mathcal{L}(\tilde{\mathcal{H}})$ by

$$\pi(y)=W_{\mathcal{H}}^*U_{\mathcal{H}}K_{\mathcal{H}}\big(PyP^*\otimes_{\mathcal{Z}}1\big)K_{\mathcal{H}}^*U_{\mathcal{H}}^*W_{\mathcal{H}},$$

for any $y \in \mathcal{M} \otimes_{\mathcal{Z}} \mathcal{Z}'_S$.

For any $a = \int_X^{\oplus} a(x) d\mu(x) \in \mathcal{M} = \int_X^{\oplus} \mathcal{M}(x) d\mu(x)$, a function

$$\gamma \in \mathcal{G}_x \longmapsto \alpha_{\gamma^{-1}}\big(a(r(\gamma))\big)$$

defines an element, denoted by $\alpha(a)(x)$, in $\mathcal{M}(x) \bar{\otimes} L^\infty(\mathcal{G}_x, \lambda_x)$. We put

$$\alpha(a) = \int_X^{\oplus} \alpha(a)(x) d\mu(x).$$

Then it is clear that $\alpha(a)$ belongs to $\mathcal{M} \otimes_{\mathcal{Z}} \mathcal{Z}'_S$. Note that $\alpha(\mathcal{M}) = \{\alpha(a) : a \in \mathcal{M}\}$ is a von Neumann algebra. $\mathcal{M} \otimes_{\mathcal{Z}} \mathcal{Z}'_S$ clearly contains $\mathbf{C} \otimes_{\mathcal{Z}} L^\infty(\mathcal{G}, \nu)$. Since the $\mathcal{Z}$ action $h \in \mathcal{Z} \longmapsto M(h \circ s)$ on $L^2(\mathcal{G}, \nu)$ commutes with the action of $\mathcal{R}(\mathcal{G})$, so that, on $\mathcal{H} \otimes_\mu L^2(\mathcal{G}, \nu)_{\mathcal{Z}}$, the relative tensor product $1 \otimes_{\mathcal{Z}} b$, $(b \in \mathcal{R}(\mathcal{G}))$ makes sense. It follows that $\mathcal{M} \otimes_{\mathcal{Z}} \mathcal{Z}'_S$ also contains the algebra $\mathbf{C} \otimes_{\mathcal{Z}} \mathcal{R}(\mathcal{G})$.

Lemma 6.1. *The von Neumann algebra generated by the sets $\alpha(\mathcal{M})$ and $\mathbf{C} \otimes_{\mathcal{Z}} L^\infty(\mathcal{G}, \nu)$ on $\mathcal{H} \otimes_\mu L^2(\mathcal{G}, \nu)_{\mathcal{Z}}$ is exactly $\mathcal{M} \otimes_{\mathcal{Z}} L^\infty(\mathcal{G}, \nu)$.*

Proof. We note first that, according to a direct integral decomposition

$$\mathcal{H} \otimes_\mu L^2(\mathcal{G}, \nu)_{\mathcal{Z}} = \int_X^{\oplus} \mathcal{H}(x) \otimes L^2(\mathcal{G}_{\mathrm{x}}, \lambda'_{\mathrm{x}}) d\mu(x),$$

$\mathcal{M} \otimes_{\mathcal{Z}} L^\infty(\mathcal{G}, \nu)$ is decomposed into

$$\mathcal{M} \otimes_{\mathcal{Z}} L^\infty(\mathcal{G}, \nu) = \int_X^{\oplus} \mathcal{M}(x) \otimes L^\infty(\mathcal{G}_{\mathrm{x}}, \lambda_{\mathrm{x}}) d\mu(x).$$

Let $\mathcal{Q} = \alpha(\mathcal{M}) \vee \{\mathbf{C} \otimes_{\mathcal{Z}} L^\infty(\mathcal{G}, \nu)\}$. Obviously, the algebra $\mathcal{Q}$ is included in $\mathcal{M} \otimes_{\mathcal{Z}} L^\infty(\mathcal{G}, \nu)$. Suppose that φ is in the predual of $\mathcal{M} \otimes_{\mathcal{Z}} L^\infty(\mathcal{G}, \nu)$ such that it vanishes on the von Neumann subalgebra $\mathcal{Q}$. From [**T3**], φ can be expressed as

$$\varphi = \int_X^{\oplus} \varphi_x d\mu(x),$$

where the family $\{\varphi_x\}_{x\in X}$ is an integrable field of normal functionals in the preduals of the algebras $\mathcal{M}(x)\bar{\otimes} L^\infty(\mathcal{G}_x, \lambda'_x)$. Moreover, each φ_x is a $\mathcal{M}(x)_*$-valued λ'_x-measurable function on $\mathcal{G}_x$ with $\int \|\varphi_x(\gamma)\| d\lambda'_x(\gamma) < \infty$. Now, for any $a = \int_X^{\oplus} a(x)d\mu(x) \in \mathcal{M} = \int_X^{\oplus} \mathcal{M}(x)d\mu(x)$, $f = \int_X^{\oplus} f_x d\mu(x) \in L^\infty(\mathcal{G}, \nu) = \int_X^{\oplus} L^\infty(\mathcal{G}_x, \lambda'_x)d\mu(x)$ and $g \in \mathcal{K}(X)$, we have, by assumption,

$$\int\int g(x) f_x(\gamma) < \alpha_{\gamma^{-1}}\big(a(r(\gamma))\big), \varphi_x(\gamma) > d\lambda'_x(\gamma)d\mu(x) = 0.$$

This implies

$$\int f_x(\gamma) < \alpha_{\gamma^{-1}}\big(a(r(\gamma))\big), \varphi_x(\gamma) > d\lambda'_x(\gamma) = 0$$

for μ a.e. $x \in X$. From this, it follows that, for μ a.e. $x \in X$, $\varphi_x = 0$. Hence $\mathcal{Q}$ must coincide with $\mathcal{M}\otimes_{\mathcal{Z}} L^\infty(\mathcal{G}, \nu)$. Q.E.D.

Next lemma shows that the algebras $L^\infty(\mathcal{G}, \nu)$ and $\mathcal{R}(\mathcal{G})$ act irreducibly on $L^2(\mathcal{G}, \nu)$.

Lemma 6.2. *On $\mathcal{H}\otimes_\mu L^2(\mathcal{G}, \nu)_{\mathcal{Z}}$, the algebras $\mathbf{C}\otimes_{\mathcal{Z}} L^\infty(\mathcal{G}, \nu)$ and $\mathbf{C}\otimes_{\mathcal{Z}}\mathcal{R}(\mathcal{G})$ generate the von Neumann algebra $\mathbf{C}\otimes_{\mathcal{Z}}\mathcal{Z}'_S$.*

Proof. Through a direct integral decomposition, $\mathcal{H}\otimes_\mu L^2(\mathcal{G}, \nu)_{\mathcal{Z}} = \int_X^{\oplus} \mathcal{H}(x) \otimes L^2(\mathcal{G}_x, \lambda'_x)\, d\mu(x)$, $\mathbf{C}\otimes_{\mathcal{Z}}\mathcal{Z}'_S$ can be written as

$$\mathbf{C}\otimes_{\mathcal{Z}}\mathcal{Z}'_S = \int_X^{\oplus} \mathbf{C}\otimes\mathcal{L}\big(L^2(\mathcal{G}_x, \lambda'_x)\big)d\mu(x).$$

Hence it suffices to show that, for any $x \in X$,

$$L^\infty(\mathcal{G}_x, \lambda'_x) \vee \{ L(f)_x : f \in \mathcal{A}_I \} = \mathcal{L}(L^2(\mathcal{G}_x, \lambda'_x)),$$

or, equivalently,

$$L^\infty(\mathcal{G}_x, \lambda'_x) \cap \{ L(f)_x : f \in \mathcal{A}_I \}' = \mathbf{C},$$

where the operator $L(f)_x$ is the x-component in the direct integral decomposition $L(f) = \int_X^\oplus L(f)_x d\mu(x)$ on $L^2(\mathcal{G}, \nu)_Z = \int_X^\oplus L^2(\mathcal{G}_x, \lambda'_x) d\mu(x)$. Note that, for $\xi \in L^2(\mathcal{G}_x, \lambda'_x)$,

$$\{L(f)_x \xi\}(\gamma) = \int f(\gamma{\gamma_1}^{-1}) \xi(\gamma_1) d\lambda_x(\gamma_1).$$

We fix $x \in X$. Suppose that $g \in L^\infty(\mathcal{G}_x, \lambda'_x) \cap \{ L(f)_x : f \in \mathcal{A}_I \}'$. If $M_x(g)$ indicates the operator of multiplication by the function g on $L^2(\mathcal{G}_x, \lambda'_x)$, then, for ξ as above, we have the identity $M_x(g) L(f)_x \xi = L(f)_x M_x(g) \xi$, which implies that

$$\int f(\gamma{\gamma_1}^{-1}) \{g(\gamma_1) - g(\gamma)\} \xi(\gamma_1) d\lambda_x(\gamma_1) = 0$$

for λ_x a.e.$\gamma \in \mathcal{G}_x$. Since $\mathcal{A}_I$ is dense, it follows from this equation that g is constant almost everywhere. Q.E.D.

Combining the preceding two lemmas, we conclude

Corollary 6.3. *The von Neumann algebra $\mathcal{M} \otimes_Z \mathcal{Z}'_S$ is generated by the algebras $\alpha(\mathcal{M})$, $\mathbf{C} \otimes_Z L^\infty(\mathcal{G}, \nu)$ and $\mathbf{C} \otimes_Z \mathcal{R}(\mathcal{G})$.*

Our next aim is to locate the image of $\mathcal{M} \otimes_Z \mathcal{Z}'_S$ under the $*$-isomorphism π defined earlier. For this, we first compute the following:

$$\{W^*_{\mathcal{H}} U_{\mathcal{H}} K_{\mathcal{H}} \big(P(1 \otimes_Z M(f)) P^* \otimes_Z 1\big)(\xi \otimes_\mu g)\}(\gamma_1, \gamma_2)$$

$$= u(\gamma_1)\{U_{\mathcal{H}}K_{\mathcal{H}}\big(P(1\otimes_Z M(f))P^*\xi\otimes_\mu g\big)\}(\gamma_1,\gamma_1{}^{-1}\gamma_2)$$

$$= \delta(\gamma_1{}^{-1}\gamma_2)^{1/2}u(\gamma_2)\{K_{\mathcal{H}}\big(P(1\otimes_Z M(f))P^*\xi\otimes_\mu g\big)\}(\gamma_2,\gamma_1{}^{-1}\gamma_2)$$

$$= \delta(\gamma_1{}^{-1}\gamma_2)^{1/2}u(\gamma_2)g(\gamma_1{}^{-1}\gamma_2)u(\gamma_2)^*\{P(1\otimes_Z M(f))P^*\xi\}(\gamma_2)$$

$$= \delta(\gamma_1{}^{-1}\gamma_2)^{1/2}g(\gamma_1{}^{-1}\gamma_2)f(\gamma_2)\xi(\gamma_2),$$

where $f \in L^\infty(\mathcal{G},\nu)$, $\xi \in \hat{\mathcal{H}}$, $g \in D(L^2(\mathcal{G},\nu)_Z,\mu)$ and $(\gamma_1,\gamma_2) \in \mathcal{H}^{(2)}$. At the same time, we have

$$\{\big(1\otimes_Z 1\otimes_Z M(f)\big)W_{\mathcal{H}}^* U_{\mathcal{H}}K_{\mathcal{H}}(\xi\otimes_\mu g)\}(\gamma_1,\gamma_2)$$

$$= f(\gamma_2)\{W_{\mathcal{H}}^* U_{\mathcal{H}}K_{\mathcal{H}}(\xi\otimes_\mu g)\}(\gamma_1,\gamma_2)$$

$$= f(\gamma_2)u(\gamma_1)\{U_{\mathcal{H}}K_{\mathcal{H}}(\xi\otimes_\mu g)\}(\gamma_1,\gamma_1{}^{-1}\gamma_2)$$

$$= f(\gamma_2)u(\gamma_1)\delta(\gamma_1{}^{-1}\gamma_2)^{1/2}\{K_{\mathcal{H}}(\xi\otimes_\mu g)\}(\gamma_2,\gamma_1{}^{-1}\gamma_2)$$

$$= \delta(\gamma_1{}^{-1}\gamma_2)^{1/2}g(\gamma_1{}^{-1}\gamma_2)f(\gamma_2)\xi(\gamma_2).$$

This shows that $\pi(1\otimes_Z M(f)) = 1\otimes_Z 1\otimes_Z M(f)$.

For any $b = \int_X^\oplus b(x)d\mu(x) \in \mathcal{M} = \int_X^\oplus \mathcal{M}(x)d\mu(x)$, $f \in \mathcal{A}_I$, $\xi \in \mathcal{H}$ and $\gamma \in \mathcal{G}$, we calculate

$$\{P\alpha(b)\big(1\otimes_Z L(f)\big)P^*\xi\}(\gamma)$$

$$= u(\gamma)\{\alpha(b)\big(1\otimes_Z L(f)\big)P^*\xi\}(\gamma)$$

$$= u(\gamma)\alpha_{\gamma^{-1}}\big(b(r(\gamma))\big)\{\big(1\otimes_Z L(f)\big)P^*\xi\}(\gamma)$$

$$= u(\gamma)\alpha_{\gamma^{-1}}\big(b(r(\gamma))\big)\int f(\gamma\gamma_1{}^{-1})\{P^*\xi\}(\gamma_1)d\lambda_{s(\gamma)}(\gamma_1)$$

$$= \int f(\gamma\gamma_1{}^{-1})b(r(\gamma))u(\gamma)u(\gamma_1)^*\xi(\gamma_1)d\lambda_{s(\gamma)}(\gamma_1)$$

$$= \int f(\gamma\gamma_1{}^{-1})b(r(\gamma))u(\gamma\gamma_1{}^{-1})\xi(\gamma_1)d\lambda_{s(\gamma)}(\gamma_1)$$

$$= \int f(\gamma_1)b(r(\gamma_1))u(\gamma_1)\xi(\gamma_1{}^{-1}\gamma)d\lambda^{r(\gamma)}(\gamma_1).$$

From this calculation, we have, with ξ, g and (γ_1,γ_2) as above,

$$\{W_{\mathcal{H}}^* U_{\mathcal{H}} K_{\mathcal{H}}\big(P\alpha(b)(1\otimes_Z L(f))P^*\otimes_Z 1\big)(\xi\otimes_\mu g)\}(\gamma_1,\gamma_2)$$

$$= u(\gamma_1)\{U_{\mathcal{H}} K_{\mathcal{H}}\big(P\alpha(b)(1\otimes_Z L(f))P^*\xi\otimes_\mu g\big)\}(\gamma_1,{\gamma_1}^{-1}\gamma_2)$$

$$= \delta({\gamma_1}^{-1}\gamma_2)^{1/2}u(\gamma_2)\{K_{\mathcal{H}}\big(P\alpha(b)(1\otimes_Z L(f))P^*\xi\otimes_\mu g\big)\}(\gamma_2,{\gamma_1}^{-1}\gamma_2)$$

$$= \delta({\gamma_1}^{-1}\gamma_2)^{1/2}u(\gamma_2)g({\gamma_1}^{-1}\gamma_2)u(\gamma_2)^*\{P\alpha(b)(1\otimes_Z L(f))P^*\xi\}(\gamma_2)$$

$$= \delta({\gamma_1}^{-1}\gamma_2)^{1/2}g({\gamma_1}^{-1}\gamma_2)\int f(\gamma_3)b(r(\gamma_3))u(\gamma_3)\xi({\gamma_3}^{-1}\gamma_2)d\lambda^{r(\gamma_2)}(\gamma_3).$$

In the meantime, with $A\in\mathcal{S}(\mathcal{M})$, $A(\gamma)=f(\gamma)b(r(\gamma))u(\gamma)\otimes\lambda(\gamma)$, we have

$$\{\hat\alpha\big(\Phi(A)\big)W_{\mathcal{H}}^* U_{\mathcal{H}} K_{\mathcal{H}}(\xi\otimes_\mu g)\}(\gamma_1,\gamma_2)$$

$$= \{\Phi'(A\otimes\lambda)W_{\mathcal{H}}^* U_{\mathcal{H}} K_{\mathcal{H}}(\xi\otimes_\mu g)\}(\gamma_1,\gamma_2)$$

$$= \int f(\gamma_3)b(r(\gamma_3))u(\gamma_3)\{W_{\mathcal{H}}^* U_{\mathcal{H}} K_{\mathcal{H}}(\xi\otimes_\mu g)\}(\gamma_3^{-1}\gamma_1,\gamma_3^{-1}\gamma_2)d\lambda^{r(\gamma_2)}(\gamma_3)$$

$$= \int f(\gamma_3)b(r(\gamma_3))u(\gamma_1)\{U_{\mathcal{H}} K_{\mathcal{H}}(\xi\otimes_\mu g)\}(\gamma_3^{-1}\gamma_1,\gamma_3^{-1}\gamma_2)d\lambda^{r(\gamma_2)}(\gamma_3)$$

$$= \int f(\gamma_3)b(r(\gamma_3))\delta(\gamma_1^{-1}\gamma_2)^{1/2}u(\gamma_2)\{K_{\mathcal{H}}(\xi\otimes_\mu g)\}(\gamma_3^{-1}\gamma_2,\gamma_1^{-1}\gamma_2)d\lambda^{r(\gamma_2)}(\gamma_3)$$

$$= \delta(\gamma_1^{-1}\gamma_2)^{1/2}\int f(\gamma_3)b(r(\gamma_3))u(\gamma_2)g(\gamma_1^{-1}\gamma_2)u(\gamma_3^{-1}\gamma_2)^*\xi(\gamma_3^{-1}\gamma_2)d\lambda^{r(\gamma_2)}(\gamma_3)$$

$$= \delta(\gamma_1^{-1}\gamma_2)^{1/2}g(\gamma_1^{-1}\gamma_2)\int f(\gamma_3)b(r(\gamma_3))u(\gamma_3)\xi(\gamma_3^{-1}\gamma_2)d\lambda^{r(\gamma_2)}(\gamma_3),$$

This proves that $\pi\big(\alpha(b)(1\otimes_Z L(f))\big)=\hat\alpha\big(\Phi(A)\big)$. But this $A\in\mathcal{S}(\mathcal{M})$ defined above is nothing other than $A_{a_b,f}$ in the notation of [**Y3**] (see the proof of Lemma 2.5 of [**Y3**]). From the proof of Theorem 2.14 of [**Y3**], we have that $\Phi(A_{a_b,f})=(b\otimes_Z 1)(u\otimes\lambda)(f)$. Hence it follows that $\pi(\alpha(b)(1\otimes_Z L(f)))=\hat\alpha((b\otimes_Z 1)(u\otimes\lambda)(f))$. From this together with combination of Corollary 6.3 and Theorem 2.14 of [**Y3**], we finally conclude that

$$\pi\big(\mathcal{M}\otimes_Z \mathcal{Z}_S'\big)=\hat\alpha(\mathcal{M}\times_\alpha\mathcal{G})\vee\{\mathbf{C}\otimes_Z L^\infty(\mathcal{G},\nu)\}$$

$$=(\mathcal{M}\times_\alpha\mathcal{G})\times_{\hat\alpha}\mathcal{G}.$$

That is, π is a $*$-isomorphism from $\mathcal{M} \otimes_{\mathcal{Z}} \mathcal{Z}'_S$ onto the double crossed product algebra $(\mathcal{M} \times_\alpha \mathcal{G}) \times_{\hat{\alpha}} \mathcal{G}$. Moreover, from the above calculations, we conclude that the morphism π sends the algebra $\{1 \otimes_{\mathcal{Z}} M(h \circ s) : h \in \mathcal{Z}\}$ onto the one $\{1 \otimes_{\mathcal{Z}} 1 \otimes_{\mathcal{Z}} M(h \circ s) : h \in \mathcal{Z}\}$, which is, as we saw earlier, a subalgebra of the center of $(\mathcal{M} \times_\alpha \mathcal{G}) \times_{\hat{\alpha}} \mathcal{G}$. Thus we have proven

Theorem 6.4. *The pairs of von Neumann algebras and their abelian subalgebras* $(\mathcal{M} \otimes_{\mathcal{Z}} \mathcal{Z}'_S, \{1 \otimes_{\mathcal{Z}} M(h \circ s) : h \in \mathcal{Z}\})$ *and* $((\mathcal{M} \times_\alpha \mathcal{G}) \times_{\hat{\alpha}} \mathcal{G}, \{1 \otimes_{\mathcal{Z}} 1 \otimes_{\mathcal{Z}} M(h \circ s) : h \in \mathcal{Z}\})$ *are isomorphic via the isomorphism* π.

The Hilbert space $\mathcal{H} \otimes_\mu L^2(\mathcal{G}, \nu)_{\mathcal{Z}}$ decomposes, with respect to the abelian algebra $\{1 \otimes_{\mathcal{Z}} M(h \circ s) : h \in \mathcal{Z}\}$, into

$$\mathcal{H} \otimes_\mu L^2(\mathcal{G}, \nu)_{\mathcal{Z}} = \int_X^{\oplus} \mathcal{H}(x) \otimes L^2(\mathcal{G}_{\mathrm{x}}, \lambda'_{\mathrm{x}})\, d\mu(x).$$

In terms of this direct integral, $\mathcal{M} \otimes_{\mathcal{Z}} \mathcal{Z}'_S$ decomposes into

$$\mathcal{M} \otimes_{\mathcal{Z}} \mathcal{Z}'_S = \int_X^{\oplus} \mathcal{M}(x) \bar{\otimes} \mathcal{L}(L^2(\mathcal{G}_x, \lambda'_x)) d\mu(x).$$

Let

$$\tilde{\mathcal{H}} = \int_X^{\oplus} \tilde{\mathcal{H}}(x) d\mu(x)$$

be the direct integral decomposition of $\tilde{\mathcal{H}}$ with respect to the abelian algebra $\{1 \otimes_{\mathcal{Z}} 1 \otimes_{\mathcal{Z}} M(h \circ s) : h \in \mathcal{Z}\}$. Let $\mathcal{P} = (\mathcal{M} \times_\alpha \mathcal{G}) \times_{\hat{\alpha}} \mathcal{G}$ and

$$\mathcal{P} = \int_X^{\oplus} \mathcal{P}(x)\, d\mu(x)$$

be its direct integral decomposition through the above decomposition of $\tilde{\mathcal{H}}$. Then, by Theorem 6.4, the isomorphism π decomposes into a measurable field $\{\pi_x\}_{x\in X}$ of $*$-isomorphisms from $\mathcal{M}(x)\bar{\otimes}\mathcal{L}(L^2(\mathcal{G}_x,\lambda'_x))$ onto $\mathcal{P}(x)$.

Let us consider the Hilbert space $\tilde{\mathcal{H}}_3 = \mathcal{H}\otimes_{\mu}{}_{\mathcal{Z}}L^2(\mathcal{G},\nu)\otimes_{\mu}L^2(\mathcal{G},\nu)_{\mathcal{Z}}$ which has a direct integral decomposition $\tilde{\mathcal{H}}_3 = \int_X^{\oplus}\mathcal{H}(x)\otimes L^2(\mathcal{G}^x,\lambda^x)\otimes L^2(\mathcal{G}_x,\lambda'_x)d\mu(x)$. As usual, we regard this Hilbert space as the set of all functions η from $\mathcal{F}^{(2)}$ into $\prod_{x\in X}\mathcal{H}(x)$ satisfying (i) $\eta(\gamma_1,\gamma_2)\in\mathcal{H}(r(\gamma_1))$, (ii) the function $x\in X\longmapsto \int\int f_{k.x}(\gamma_1)\,\tilde{f}_{l,x}(\gamma_2)(\,\xi_{m,x}\mid \eta(\gamma_1,\gamma_2)\,)d\lambda^x(\gamma_1)d\lambda'_x(\gamma_2)$ is μ-measurable for any k, l and $m\in\mathbf{N}$, (iii) The quantity $\int\|\eta(\gamma_1,\gamma_2)\|^2d\nu_3(\gamma_1,\gamma_2)$ is bounded. Then the norm of a vector η is, of course, given by $\|\eta\|=\left(\int\|\eta(\gamma_1,\gamma_2)\|^2d\nu_3(\gamma_1,\gamma_2)\right)^{1/2}$. Under this identification, we can join Hilbert spaces $\tilde{\mathcal{H}}$ and $\tilde{\mathcal{H}}_3$ by a unitary $R:\tilde{\mathcal{H}}\longrightarrow\tilde{\mathcal{H}}_3$ defined by $\{R\xi\}(\gamma_1,\gamma_2) = u(\gamma_2)^*\xi(\gamma_2\gamma_1,\gamma_2)$, $(\,\xi\in\tilde{\mathcal{H}},\ (\gamma_1,\gamma_2)\in\mathcal{F}^{(2)}\,)$. It is easily verified that R is indeed a unitary with its adjoint given by $\{R^*\eta\}(\gamma_1,\gamma_2)=u(\gamma_2)\eta(\gamma_2{}^{-1}\gamma_1,\gamma_2)$, $(\,\eta\in\tilde{\mathcal{H}}_3,\ (\gamma_1,\gamma_2)\in\mathcal{H}^{(2)}\,)$. Then we observe the following:

$$\begin{aligned}&\{R\big(1\otimes_{\mathcal{Z}}1\otimes_{\mathcal{Z}}M(h\circ s)\big)R^*\eta\}(\gamma_1,\gamma_2)\\&=u(\gamma_2)^*\{\big(1\otimes_{\mathcal{Z}}1\otimes_{\mathcal{Z}}M(h\circ s)\big)R^*\eta\}(\gamma_2\gamma_1,\gamma_2)\\&=h(s(\gamma_2))u(\gamma_2)^*\{R^*\eta\}(\gamma_2\gamma_1,\gamma_2)\\&=h(s(\gamma_2))\eta(\gamma_1,\gamma_2)\\&=\{\big(1\otimes_{\mathcal{Z}}1\otimes_{\mathcal{Z}}M(h\circ s)\big)\eta\}(\gamma_1,\gamma_2),\end{aligned}$$

whenever $h\in\mathcal{Z}$ and $\eta\in\tilde{\mathcal{H}}_3$. The $\mathcal{Z}$ action appeared above at the end is the canonical $\mathcal{Z}$ action on $\tilde{\mathcal{H}}_3$. It follows that R is a decomposable operator with decomposition $\{R(x)\}_{x\in X}$, where $R(x)$ is a unitary operator from $\tilde{\mathcal{H}}(x)$ onto $\tilde{\mathcal{H}}_3(x) = \mathcal{H}(x)\otimes$

$L^2(\mathcal{G}^x, \lambda^x) \otimes L^2(\mathcal{G}_x, \lambda'_x)$. Note that, by the result of § 5, we have

$$\tilde{\mathcal{H}}(x) = \int_{[x]}^{\oplus} L^2(\mathcal{G}_x^y, \tau_{x,y}) \otimes \mathcal{H}(y) \otimes L^2(\mathcal{G}^y, \lambda^y) d\tilde{\nu}_x(y).$$

Refer to § 5 for the notations $\tilde{\nu}_x$, $\tau_{x,y}$ and so forth. Then every element ζ in $\tilde{\mathcal{H}}(x)$ is a certain kind of functions with two variables such that $\zeta(\ \ ,\gamma_2) \in \mathcal{H}(r(\gamma_2)) \otimes L^2(\mathcal{G}^{r(\gamma_2)}, \lambda^{r(\gamma_2)})$, $(\gamma_2 \in \mathcal{G}_x)$, satisfying other suitable conditions. Then $R(x)$ can be described more explicitly as follows:

$$\{R(x)\,\zeta\}(\gamma_1, \gamma_2) = u(\gamma_2)^* \zeta(\gamma_2\gamma_1, \gamma_2), \qquad ((\gamma_1, \gamma_2) \in \mathcal{G}^x \times \mathcal{G}_x).$$

and

$$\{R(x)^*\,\eta\}(\gamma_1, \gamma_2) = u(\gamma_2)\eta({\gamma_2}^{-1}\gamma_1, \gamma_2).$$

Recall that, by the consequence of § 5, for each $\gamma \in \mathcal{G}$, we have a unitary $\rho_{\tilde{\mathcal{H}}}(\gamma)$ from $\tilde{\mathcal{H}}(s(\gamma))$ onto $\tilde{\mathcal{H}}(r(\gamma))$ which implements the bidual action $\tilde{\alpha}_\gamma$. Then we calculate

$$\begin{aligned}
&\{R(r(\gamma))\rho_{\tilde{\mathcal{H}}}(\gamma)R(s(\gamma))^*\eta\}(\gamma_1, \gamma_2) \\
&= u(\gamma_2)^*\{\rho_{\tilde{\mathcal{H}}}(\gamma)R(s(\gamma))^*\eta\}(\gamma_2\gamma_1, \gamma_2) \\
&= u(\gamma_2)^*\delta(\gamma)^{1/2}\{R(s(\gamma))^*\eta\}(\gamma_2\gamma_1, \gamma_2\gamma) \\
&= \delta(\gamma)^{1/2}u(\gamma_2)^*u(\gamma_2\gamma)\eta(\gamma^{-1}\gamma_1, \gamma_2\gamma) \\
&= \delta(\gamma)^{1/2}u(\gamma)\eta(\gamma^{-1}\gamma_1, \gamma_2\gamma) \\
&= \{(u(\gamma) \otimes \lambda(\gamma) \otimes \rho(\gamma))\eta\}(\gamma_1, \gamma_2),
\end{aligned}$$

for any $\eta \in \tilde{\mathcal{H}}_3(s(\gamma))$. Thus we obtain $R(r(\gamma))\rho_{\tilde{\mathcal{H}}}(\gamma)R(s(\gamma))^* = u(\gamma) \otimes \lambda(\gamma) \otimes \rho(\gamma)$. Moreover, an easy, but long calculation yields

$$R(x)\pi_x(\alpha(a)(x))R(x)^* = \alpha(a)(x)_{13}$$

$$R(x)\pi_x(1 \otimes L(f)_x)R(x)^* = 1 \otimes 1 \otimes L(f)_x$$

$$R(x)\pi_x(1 \otimes M_x(g))R(x)^* = 1 \otimes 1 \otimes M_x(g),$$

where $a \in \mathcal{M}$, $f \in \mathcal{A}_I$ and $g \in L^\infty(\mathcal{G}, \nu)$. Here Y_{13} stands for the operator obtained by the following manner from Y: if Y is an operator on $\mathcal{H}_1 \otimes \mathcal{H}_2$ and $\mathcal{H}_3$ is another Hilbert space, then the operator Y_{13} on $\mathcal{H}_1 \otimes \mathcal{H}_3 \otimes \mathcal{H}_2$ is given by

$$Y_{13} = (1 \otimes \sigma^*)(Y \otimes 1_{\mathcal{H}_3})(1 \otimes \sigma),$$

where σ is a unitary from $\mathcal{H}_3 \otimes \mathcal{H}_2$ onto $\mathcal{H}_2 \otimes \mathcal{H}_3$ defined by $\sigma(\xi \otimes \eta) = \eta \otimes \xi$.

Form what we discussed above, it follows that the bidual action $(\mathcal{G}, \{\mathcal{P}(x), \tilde{\mathcal{H}}(x)\}, \{\tilde{\alpha}_\gamma = \mathrm{Ad}\,\rho_{\tilde{\mathcal{H}}}(\gamma)\})$ is conjugate to a new action $(\mathcal{G}, \{\mathcal{M}(x) \bar{\otimes} \mathcal{L}(L^2(\mathcal{G}_x, \lambda'_x)), \mathcal{H}(x) \otimes L^2(\mathcal{G}_x, \lambda'_x)\}, \{\mathrm{Ad}\,u(\gamma) \otimes \rho(\gamma)\})$ via the measurable field $\{\pi_x\}$ of $*$-isomorphisms from $\mathcal{M}(x) \bar{\otimes} \mathcal{L}(L^2(\mathcal{G}_x, \lambda'_x))$ onto $\mathcal{P}(x)$. Thus we have proven the main theorem in this section:

Theorem 6.5. **(Duality for actions)** *Suppose that* $(\mathcal{G}, \{\mathcal{M}(x), \mathcal{H}(x)\}, \{\alpha_\gamma = \mathrm{Ad}\,u(\gamma)\})$ *is an action of* $\mathcal{G}$. *Then the bidual action of the original one is conjugate to the action* $(\mathcal{G}, \{\mathcal{M}(x) \bar{\otimes} \mathcal{L}(L^2(\mathcal{G}_x, \lambda'_x)), \mathcal{H}(x) \otimes L^2(\mathcal{G}_x, \lambda'_x)\}, \{\mathrm{Ad}\,u(\gamma) \otimes \rho(\gamma)\})$.

As a corollary of the above theorem, we may obtain Takesaki duality for actions of locally compact (second countable) groups on von Neumann algebras.

§ 7. Duality for integrable coactions on von Neumann algebras

In § 6, we proved duality for actions of measured groupoids on von Neumann algebras, which may be regarded as as extension of Takesaki duality from group actions to groupoid actions. In this section, we shall devote ourself to the same problem for coactions of measured groupoids. However, in order to have much control on a given coaction, we have to put a certain condition on it. The condition we will impose is called "integrability". Under this assumption, we shall show the duality.

Let $\{\mathcal{H}(x) = \mathcal{H}\}_{x\in X}$ be a constant field of Hilbert spaces over X. Suppose that we are given a family $\{\mathcal{N}(\gamma)\}_{\gamma\in\mathcal{G}}$ of weakly closed subspaces such that (1) $\mathcal{N}(\gamma) \subseteq \mathcal{L}(\mathcal{H}(s(\gamma)), \mathcal{H}(r(\gamma)))$ for every $\gamma \in \mathcal{G}$, (2) $\mathcal{N}(\gamma_1)\mathcal{N}(\gamma_2) \subseteq \mathcal{N}(\gamma_1\gamma_2)$ whenever $(\gamma_1, \gamma_2) \in \mathcal{G}^{(2)}$, (3) $\mathcal{N}(\gamma)^* = \mathcal{N}(\gamma^{-1})$ for any $\gamma \in \mathcal{G}$. Note that, because of (2) and (3), each $\mathcal{N}(x)$ $(x \in X)$ is a von Neumann algebra on $\mathcal{H}(x)$. As we constructed the crossed product algebra from a given action in § 4, we define the set $\mathcal{S}(\mathcal{G}, \prod_{\gamma\in\mathcal{G}}\mathcal{N}(\gamma))$ of all sections from $\mathcal{G}$ into $\prod_{\gamma\in\mathcal{G}}\mathcal{N}(\gamma)$. Namely, every section B in $\mathcal{S}(\mathcal{G}, \prod_{\gamma\in\mathcal{G}}\mathcal{N}(\gamma))$ is one with properties:

(1) The function $\gamma \in \mathcal{G} \mapsto (B(\gamma)\xi \mid \eta)$ is measurable for any ξ and η in $\mathcal{H}$;

(2) $\|B\| = \max\{ \|\lambda(\|B(\cdot)\|)\|_\infty, \|\lambda(\|B^\sharp(\cdot)\|)\|_\infty \}$ is finite, where $B^\sharp(\gamma) = \delta(\gamma)^{-1}B(\gamma^{-1})^*$.

We shall write, as usual, $\mathcal{S}(\mathcal{N})$ for the above set for short. We can turn this set into a $\sharp$-algebra in the similar way we did $\mathcal{S}(\mathcal{M})$ in § 4. For each B, we define a section B_λ from $\mathcal{G}$ into $\prod_{\gamma\in\mathcal{G}}\mathcal{N}(\gamma)\otimes\lambda(\gamma)$ by $B_\lambda(\gamma) = B(\gamma)\otimes\lambda(\gamma)$. Thus every B_λ is a bounded operator from $\mathcal{H}(s(\gamma))\otimes L^2(\mathcal{G}^{s(\gamma)},\lambda^{s(\gamma)})$ into $\mathcal{H}(r(\gamma))\otimes L^2(\mathcal{G}^{r(\gamma)},\lambda^{r(\gamma)})$. We may define a (convolution) product of two elements B_λ, C_λ ($B, C \in \mathcal{S}(\mathcal{N})$, and a $\sharp$-operation of B_λ in an obvious manner. Then we have $B_\lambda * C_\lambda(\gamma) = B*C(\gamma)\otimes\lambda(\gamma)$ and $B_\lambda^\sharp(\gamma) = B^\sharp(\gamma)\otimes\lambda(\gamma)$.

Next we consider a representation Φ_0 of the algebra $\mathcal{S}(\mathcal{N})$ on a Hilbert space $\mathcal{H}'\otimes_{\mu\mathcal{Z}}L^2(\mathcal{G},\nu)$, where $\mathcal{H}' = \int_X^\oplus \mathcal{H}(x)d\mu(x) = \mathcal{H}\otimes L^2(X,\mu)$. It is clear that the above relative tensor product is just an ordinary tensor product $\mathcal{H}\otimes L^2(\mathcal{G},\nu)$. Hence every element in the above relative tensor product may be viewed as an $\mathcal{H}$-valued square integrable function on $\mathcal{G}$. We denote by $\mathcal{K}$ the Hilbert space $\mathcal{H}'\otimes_{\mu\mathcal{Z}}L^2(\mathcal{G},\nu) = \mathcal{H}\otimes L^2(\mathcal{G},\nu)$. We then define an operator $\Phi_0(B_\lambda)$ on $\mathcal{K}$ by

$$\{\Phi_0(B_\lambda)\xi\}(\gamma) = \int B(\gamma_1)\xi({\gamma_1}^{-1}\gamma)d\lambda^{r(\gamma)}(\gamma_1)$$

for any $\xi \in \mathcal{K}$. As in § 4, one can prove that Φ_0 is a nondegenerate $*$-representation on $\mathcal{K}$. We let $\mathcal{N}$ denote the von Neumann algebra generated by this representation. It is easy to see that $\mathcal{Z}$ on $\mathcal{K}$ giving the canonical $\mathcal{Z}$ action is contained in $\mathcal{N}$. Next we consider the Hilbert space $\mathcal{H}'\otimes_{\mu\mathcal{Z}}L^2(\mathcal{G},\nu)\otimes_{\mu\mathcal{Z}}L^2(\mathcal{G},\nu)$. It is easily verified that this space is (isomorphic to) $\mathcal{H}\otimes\big({}_{\mathcal{Z}}L^2(\mathcal{G},\nu)\otimes_{\mu\mathcal{Z}}L^2(\mathcal{G},\nu)\big) = \mathcal{H}\otimes L^2(\mathcal{H}^{(2)},\nu_1)$. On $\mathcal{K}$, besides the canonical $\mathcal{Z}$ action, we have another $\mathcal{Z}$-module structure: $h \in \mathcal{Z} \mapsto 1\otimes M(h\circ s)$. We shall write $\mathcal{K}_{\mathcal{Z}}$ for $\mathcal{K}$ if this $\mathcal{Z}$ action on $\mathcal{K}$ is particularly under consideration. We look at the Hilbert space $\mathcal{K}_{\mathcal{Z}}\otimes_{\mu\mathcal{Z}}L^2(\mathcal{G},\nu)$. Let $\xi_i \in \mathcal{H}$, $f_i \in L^2(\mathcal{G},\nu)$ and $g_i \in D({}_{\mathcal{Z}}L^2(\mathcal{G},\nu),\mu)$ ($i = 1,2$). Then the inner product of $(\xi_1\otimes f_1)\otimes_\mu g_1$ and $(\xi_2\otimes f_2)\otimes_\mu g_2$ on $\mathcal{K}_{\mathcal{Z}}\otimes_{\mu\mathcal{Z}}L^2(\mathcal{G},\nu)$ is calculated as follows:

$$\begin{aligned}&\big((\xi_1\otimes f_1)\otimes_\mu g_1 \mid (\xi_2\otimes f_2)\otimes_\mu g_2\big)\\ &= \big((1\otimes M(< g_1, g_2 > \circ s))(\xi_1\otimes f_1) \mid \xi_2\otimes f_2\big)\end{aligned}$$

$$= (\xi_1 \mid \xi_2)(M(< g_1, g_2 > \circ s) f_1 \mid f_2)$$

$$= (\xi_1 \mid \xi_2)(f_1 \otimes_\mu g_1 \mid f_2 \otimes_\mu g_2)$$

$$= (\xi_1 \otimes (f_1 \otimes_\mu g_1) \mid \xi_2 \otimes (f_2 \otimes_\mu g_2)).$$

Here the vectors $f_1 \otimes_\mu g_1$ and $f_2 \otimes_\mu g_2$ that appeared in the last equality are tensors in $L^2(\mathcal{G},\nu)_{\mathcal{Z}} \otimes_{\mu \mathcal{Z}} L^2(\mathcal{G},\nu) = L^2(\mathcal{G}^{(2)},\nu_2)$. This shows that the equation

$$V_0(\xi_1 \otimes f_1) \otimes_\mu g_1 = \xi_1 \otimes (f_1 \otimes_\mu g_1)$$

defines a unitary from $\mathcal{K}_{\mathcal{Z}} \otimes_{\mu \mathcal{Z}} L^2(\mathcal{G},\nu)$ onto $\mathcal{H} \otimes (L^2(\mathcal{G},\nu)_{\mathcal{Z}} \otimes_{\mu \mathcal{Z}} L^2(\mathcal{G},\nu))$ $= \mathcal{H} \otimes L^2(\mathcal{G}^{(2)},\nu_2)$.

On $\mathcal{K}$, every operator $\Phi_0(B_\lambda)$ $(B \in \mathcal{S}(\mathcal{N}))$ commutes with the above $\mathcal{Z}$ action. In fact, if $h \in \mathcal{Z}$ and $\xi \in \mathcal{K}$, then we calculate

$$\{\Phi_0(B_\lambda)(1 \otimes M(h \circ s))\xi\}(\gamma)$$

$$= \int B(\gamma_1)\{(1 \otimes M(h \circ s))\xi\}(\gamma_1^{-1}\gamma) d\lambda^{r(\gamma)}(\gamma_1)$$

$$= \int h(s(\gamma)) B(\gamma_1)\xi(\gamma_1^{-1}\gamma) d\lambda^{r(\gamma)}(\gamma_1)$$

$$= h(s(\gamma))\{\Phi_0(B_\lambda)\xi\}(\gamma)$$

$$= \{(1 \otimes M(h \circ s))\Phi_0(B_\lambda)\xi\}(\gamma).$$

Thus $\Phi_0(B_\lambda)(1 \otimes M(h \circ s)) = (1 \otimes M(h \circ s))\Phi_0(B_\lambda)$. From this, it follows that, on $\mathcal{K}_{\mathcal{Z}} \otimes_{\mu \mathcal{Z}} L^2(\mathcal{G},\nu)$, an operator $y \otimes_{\mathcal{Z}} 1$ makes sense for any $y \in \mathcal{N}$. Next we define a *-isomorphism δ from $\mathcal{N}$ into $\mathcal{L}(\mathcal{K} \otimes_{\mu \mathcal{Z}} L^2(\mathcal{G},\nu))$. Before we do this, recall that $\mathcal{K} \otimes_{\mu \mathcal{Z}} L^2(\mathcal{G},\nu) = \mathcal{H} \otimes L^2(\mathcal{G},\nu)$. With this in mind, δ is defined by

$$\delta(y) = (1 \otimes W^*) V_0(y \otimes_{\mathcal{Z}} 1) V_0^*(1 \otimes W) \qquad (y \in \mathcal{N}),$$

where W denotes the unitary operator from $L^2(\mathcal{H}^{(2)},\nu_1)$ onto $L^2(\mathcal{G}^{(2)},\nu_2)$ constructed in § 2. We would like to locate the image of $\mathcal{N}$ under the morphism δ. For this purpose, we look at $\delta(y)$ when y is of the form $y = \Phi_0(B_\lambda)$ $(B \in \mathcal{S}(\mathcal{N}))$. But, arguing like Lemma 4.4, we may conclude

$$\delta\big(\Phi_0(B_\lambda)\big) = \Phi_0'(B_\lambda \otimes \lambda).$$

Here the operator $\Phi_0'(B_\lambda \otimes \lambda)$ should be appropriately understood.

If T is an operator on $\mathcal{K}$ with $T \in \mathcal{Z}'$, then, since it commutes with the canonical $\mathcal{Z}$ action on $\mathcal{K}: h \in \mathcal{Z} \mapsto 1 \otimes M(h \circ s)$, T is decomposed into $T = \int_X^\oplus T(x) d\mu(x)$ according to the decomposition $\mathcal{K} = \int_X^\oplus \mathcal{H}(x) \otimes L^2(\mathcal{G}^x, \lambda^x) d\mu(x)$. Then it should be noticed that $T \in \mathcal{N}'$ if and only if $\{T(x)\}_{x \in X}$ satisfies

$$T(r(\gamma))\big(B(\gamma) \otimes \lambda(\gamma)\big) = \big(B(\gamma) \otimes \lambda(\gamma)\big)T(s(\gamma))$$

for every $B \in \mathcal{S}(\mathcal{N})$ and ν-a.e.$\gamma \in \mathcal{G}$. from this, it immediately follows that $\delta\big(\Phi_0(B_\lambda)\big) \in \mathcal{N} *_{\mathcal{Z}} \mathcal{R}(\mathcal{G})$. In fact, it can be furthrer proven that δ is actually a coaction of $\mathcal{G}$ on $\mathcal{N}$. Namely, δ satisfies the identity: $(\delta *_{\mathcal{Z}} \iota) \circ \delta = (\iota *_{\mathcal{Z}} \Gamma) \circ \delta$. The proof for this assertion exactly follows the line of arguments in § 4 in which $\hat{\alpha}$ was proven to be a coaction of $\mathcal{G}$ on the crossed product algebra. Now we are in a position to state the definition of integrability of a coaction.

Definition 7.1. A coaction $(\mathcal{G}, \mathcal{N}, \delta, \mathcal{K})$ is said to be integrable if it is constructed by the way indicated above from a system $(\{\mathcal{N}(\gamma)\}_{\gamma \in \mathcal{G}}, \{\mathcal{H}(x) = \mathcal{H}\}_{x \in X})$.

In the rest of this section, our aim is to show duality for an integrable coaction of $\mathcal{G}$ on a von Neumann algebra. So we fix an integrable coaction $(\mathcal{G}, \mathcal{N}, \delta, \mathcal{K})$ derived from $(\{\mathcal{N}(\gamma)\}, \{\mathcal{H}(x) = \mathcal{H}\})$. First we prove the following lemma.

Lemma 7.2. *On* $\mathcal{K}\otimes_{\mu\mathcal{Z}}L^2(\mathcal{G},\nu)$, $\mathcal{N}*_{\mathcal{Z}}\mathcal{L}(_{\mathcal{Z}}L^2(\mathcal{G},\nu))$ *is generated by the algebras* $\delta(\mathcal{N})$ *and* $\mathcal{Z}*_{\mathcal{Z}}\mathcal{L}(_{\mathcal{Z}}L^2(\mathcal{G},\nu))$.

Proof. It is clear that $\delta(\mathcal{N})\vee\mathcal{Z}*_{\mathcal{Z}}\mathcal{L}(_{\mathcal{Z}}L^2(\mathcal{G},\nu))$ is contained in $\mathcal{N}*_{\mathcal{Z}}\mathcal{L}(_{\mathcal{Z}}L^2(\mathcal{G},\nu))$. The reverse inclusion is equivalent to

$$\delta(\mathcal{N})'\wedge\{\mathcal{Z}'\otimes_{\mathcal{Z}}\mathbf{C}\}\subseteq\mathcal{N}'\otimes_{\mathcal{Z}}\mathbf{C}.$$

So let Q be an operator on $\mathcal{K}$ such that $Q\in\mathcal{Z}'$ and $\big[Q\otimes_{\mathcal{Z}}1,\,\delta(y)\big]=0$ for all $y\in\mathcal{N}$, where $\big[a,\,b\big]=ab-ba$. Let $Q=\int_X^{\oplus}Q(x)d\mu(x)$ be a decomposition of Q according to the direct integral $\mathcal{K}=\int_X^{\oplus}\mathcal{H}(x)\otimes L^2(\mathcal{G}^x,\,\lambda^x)d\mu(x)$. Then the condition $\big[Q\otimes_{\mathcal{Z}}1,\,\delta\big(\Phi_0(B_\lambda)\big)\big]=0\ \ (B\in\mathcal{S}(\mathcal{N}))$ implies that

$$\begin{aligned}&\int\{Q(r(\gamma))\big(B(\gamma)\otimes\lambda(\gamma)\big)\xi\}(\gamma_1)g(\gamma^{-1}\gamma_1)d\lambda^{r(\gamma_1)}(\gamma)\\&=\int\{\big(B(\gamma)\otimes\lambda(\gamma)\big)Q(s(\gamma))\xi\}(\gamma_1)g(\gamma^{-1}\gamma_1)d\lambda^{r(\gamma_1)}(\gamma)\end{aligned}$$

for ν-a.e. $\gamma_1\in\mathcal{G}$, any $\xi\in\mathcal{K}$ and $g\in D(_{\mathcal{Z}}L^2(\mathcal{G},\nu),\mu)$. From this, it follows that

$$Q(r(\gamma))\big(B(\gamma)\otimes\lambda(\gamma)\big)=\big(B(\gamma)\otimes\lambda(\gamma)\big)Q(s(\gamma))$$

for any $B\in\mathcal{S}(\mathcal{N})$ and ν-a.e.$\gamma\in\mathcal{G}$. Thus $Q\in\mathcal{N}'$. Q.E.D.

By Proposition 1.1.(2), we have that $\mathcal{Z}*_{\mathcal{Z}}\mathcal{L}(_{\mathcal{Z}}L^2(\mathcal{G},\nu))=\mathcal{Z}\otimes_{\mathcal{Z}}\mathcal{Z}'_R$. In the meantime, it can be shown by an argument similar to that in Lemma 6.2 that $\mathcal{Z}'_R=\mathcal{R}(\mathcal{G})'\vee L^\infty(\mathcal{G},\nu)$. Hence, by virtue of Lemma 7.2, we may conclude

Proposition 7.3. *The von Neumann algebra* $\mathcal{N}*_{\mathcal{Z}}\mathcal{L}(_{\mathcal{Z}}L^2(\mathcal{G},\nu))$ *is generated by the algebras* $\delta(\mathcal{N})$, $\mathbf{C}\otimes_{\mathcal{Z}}\mathcal{R}(\mathcal{G})'$ *and* $\mathbf{C}\otimes_{\mathcal{Z}}L^\infty(\mathcal{G},\nu)$.

We will next show that $\mathcal{N} *_{\mathcal{Z}} \mathcal{L}({}_{\mathcal{Z}}L^2(\mathcal{G},\nu))$ is isomorphic to the double crossed product algebra $(\mathcal{N} \times_\delta \mathcal{G}) \times_{\hat{\delta}} \mathcal{G}$. For this, we first observe the following:

$$\begin{aligned}(\delta *_{\mathcal{Z}} \iota)(1_{\mathcal{K}} \otimes_{\mathcal{Z}} T) &= 1_{\mathcal{K}} \otimes_{\mathcal{Z}} 1_{L^2(\mathcal{G},\nu)} \otimes_{\mathcal{Z}} T \\ (\delta *_{\mathcal{Z}} \iota)(1_{\mathcal{K}} \otimes_{\mathcal{Z}} M(f)) &= 1_{\mathcal{K}} \otimes_{\mathcal{Z}} 1_{L^2(\mathcal{G},\nu)} \otimes_{\mathcal{Z}} M(f) \\ (\delta *_{\mathcal{Z}} \iota)\big(\delta(\Phi_0(B_\lambda))\big) &= \Phi_0''(B_\lambda \otimes \lambda \otimes \lambda)\end{aligned} \tag{7.4}$$

for any $T \in \mathcal{R}(\mathcal{G})'$, $f \in L^\infty(\mathcal{G},\nu)$ and $B \in \mathcal{S}(\mathcal{N})$. Here $\Phi_0''(B_\lambda \otimes \lambda \otimes \lambda)$ is defined similarly when we defined $\Phi''(A \otimes \lambda \otimes \lambda)$ for $A \in \mathcal{S}(\mathcal{M})$ in § 4.

Recall that the crossed product algebra $\mathcal{N} \times_\delta \mathcal{G}$ is, by definition, generated by $\delta(\mathcal{N})$ and $\mathbf{C}_{\mathcal{K}} \otimes_{\mathcal{Z}} L^\infty(\mathcal{G},\nu)$. From an argument similar to the one we made before Theorem 6.5, it follows that the dual action is obtained by decomposing the von Neumann algebra $(1 \otimes V)(\mathcal{N} \times_\delta \mathcal{G})(1 \otimes V)^*$ on $\mathcal{H} \otimes \big({}_{\mathcal{Z}}L^2(\mathcal{G},\nu) \otimes_\mu L^2(\mathcal{G},\nu)_{\mathcal{Z}}\big)$ with respect to $\mathcal{Z}$ imbedded into the center of $(1 \otimes V)(\mathcal{N} \times_\delta \mathcal{G})(1 \otimes V)^*$ via $h \in \mathcal{Z} \mapsto 1 \otimes 1 \otimes_{\mathcal{Z}} M(h \circ s)$, where V is a unitary from ${}_{\mathcal{Z}}L^2(\mathcal{G},\nu) \otimes_\mu {}_{\mathcal{Z}}L^2(\mathcal{G},\nu) = L^2(\mathcal{H}^{(2)},\nu_1)$ onto ${}_{\mathcal{Z}}L^2(\mathcal{G},\nu) \otimes_\mu L^2(\mathcal{G},\nu)_{\mathcal{Z}} = L^2(\mathcal{F}^{(2)},\nu_3)$ defined in § 2. Then $\hat{\delta}_\gamma$ is given by $\hat{\delta}_\gamma = \mathrm{Ad}\big(1 \otimes \lambda(\gamma) \otimes \rho(\gamma)\big)$.

Let $\eta \in \mathcal{H} \otimes L^2(\mathcal{F}^{(2)},\nu_3) = \mathcal{K} \otimes_\mu L^2(\mathcal{G},\nu)_{\mathcal{Z}}$ and $B \in \mathcal{S}(\mathcal{N})$. Then

$$\begin{aligned}&\{(1 \otimes V)\delta\big(\Phi_0(B_\lambda)\big)(1 \otimes V)^* \eta\}(\gamma_1, \gamma_2) \\ &= \{(1 \otimes V)\Phi_0'(B_\lambda \otimes \lambda)(1 \otimes V)^* \eta\}(\gamma_1, \gamma_2) \\ &= \{\Phi_0'(B_\lambda \otimes \lambda)(1 \otimes V^*)\eta\}(\gamma_2\gamma_1, \gamma_2) \\ &= \int B(\gamma)\{(1 \otimes V^*)\eta\}(\gamma^{-1}\gamma_2\gamma_1, \gamma^{-1}\gamma_2)\, d\lambda^{r(\gamma_2)}(\gamma) \\ &= \int B(\gamma)\eta(\gamma_1, \gamma^{-1}\gamma_2)\, d\lambda^{r(\gamma_2)}(\gamma).\end{aligned}$$

With $f \in L^\infty(\mathcal{G},\nu)$, we also compute

$$\begin{aligned}&\{(1 \otimes V)(1 \otimes 1 \otimes_{\mathcal{Z}} M(f))(1 \otimes V^*)\eta\}(\gamma_1, \gamma_2)\\ &= \{(1 \otimes 1 \otimes_{\mathcal{Z}} M(f))(1 \otimes V^*)\eta\}(\gamma_2\gamma_1, \gamma_2)\\ &= f(\gamma_2)\{(1 \otimes V^*)\eta\}(\gamma_2\gamma_1, \gamma_2)\\ &= f(\gamma_2)\eta(\gamma_1, \gamma_2)\\ &= \{(1 \otimes 1 \otimes_{\mathcal{Z}} M(f))\eta\}(\gamma_1, \gamma_2).\end{aligned}$$

Now let $\Theta(B) = (1 \otimes V)\delta(\Phi_0(B_\lambda))(1 \otimes V^*)$ and

$$\Theta(B) = \int_X^\oplus \Theta(B)_x d\mu(x)$$

be a direct integral decomposition of $\Theta(B)$ with respect to the $\mathcal{Z}$ action mentioned above. Suppose $q \in \mathcal{A}_I$. If $A_{B,q}(\gamma) = q(\gamma)\Theta(B)_{r(\gamma)}(1 \otimes\lambda(\gamma)\otimes\rho(\gamma))\otimes\lambda(\gamma)$ $(\gamma \in \mathcal{G})$, then $A_{B,q} \in \mathcal{S}(\mathcal{N}\times_\delta\mathcal{G})$. For such a q and $f \in L^\infty(\mathcal{G},\nu)$, define $A_{f,q}(\gamma) = q(\gamma)(1 \otimes 1 \otimes M_{r(\gamma)}(f))(1 \otimes \lambda(\gamma) \otimes \rho(\gamma)) \otimes \lambda(\gamma)$ $(\gamma \in \mathcal{G})$, where $\{M_x(f)\}_{x\in X}$ is a decomposition of $M(f)$ on $L^2(\mathcal{G},\nu)_{\mathcal{Z}} = \int_X^\oplus L^2(\mathcal{G}_x,\lambda'_x)d\mu(x)$. Then $A_{f,q}$ also belongs to $\mathcal{S}(\mathcal{N}\times_\delta\mathcal{G})$. Then the double crossed product algebra $(\mathcal{N} \times_\delta \mathcal{G})\times_{\hat{\delta}} \mathcal{G}$ is generated on the Hilbert space $\hat{\mathcal{K}}=\mathcal{K} \otimes_\mu L^2(\mathcal{G},\nu)_{\mathcal{Z}} \otimes_{\mu\mathcal{Z}} L^2(\mathcal{G},\nu)$ by elements of the forms $\Phi(A_{B,q})$ and $\Phi(A_{f,q})$. Here Φ is the non-degenerate $*$-representation of $\mathcal{S}(\mathcal{N}\times_\delta\mathcal{G})$. Note that $\hat{\mathcal{K}}=\mathcal{H} \otimes ({}_{\mathcal{Z}}L^2(\mathcal{G},\nu)\otimes_\mu L^2(\mathcal{G},\nu)_{\mathcal{Z}}\otimes_\mu {}_{\mathcal{Z}}L^2(\mathcal{G},\nu))$. Now let us consider a unitary U_0 from $\mathcal{H}_1 = {}_{\mathcal{Z}}L^2(\mathcal{G},\nu)\otimes_{\mu\mathcal{Z}} L^2(\mathcal{G},\nu)\otimes_\mu {}_{\mathcal{Z}}L^2(\mathcal{G},\nu)$ onto $\mathcal{H}_2 = {}_{\mathcal{Z}}L^2(\mathcal{G},\nu)\otimes_\mu L^2(\mathcal{G},\nu)_{\mathcal{Z}}\otimes_{\mu\mathcal{Z}} L^2(\mathcal{G},\nu)$ by

$$\{U_0\xi\}(\gamma_1,\gamma_2,\gamma_3) = \xi(\gamma_2\gamma_1,\gamma_2\gamma_3,\gamma_2)$$

for any $\xi \in \mathcal{H}_1$ and $(\gamma_1,\gamma_2,\gamma_3)$ with $r(\gamma_1) = s(\gamma_2) = r(\gamma_3)$. By Fubini's theorem, we have

$$\iiiint |\xi(\gamma_2\gamma_1, \gamma_2\gamma_3, \gamma_2)|^2 d\lambda^x(\gamma_3) d\lambda'_x(\gamma_2) d\lambda^x(\gamma_1) d\mu(x)$$
$$= \iiiint |\xi(\gamma_2\gamma_1, \gamma_2\gamma_3, \gamma_2)|^2 d\lambda^{s(\gamma_2)}(\gamma_3) d\lambda'_x(\gamma_2) d\lambda^x(\gamma_1) d\mu(x)$$
$$= \iiiint |\xi(\gamma_2\gamma_1, \gamma_3, \gamma_2)|^2 d\lambda^{r(\gamma_2)}(\gamma_3) d\lambda'_x(\gamma_2) d\lambda^x(\gamma_1) d\mu(x)$$
$$= \iiiint |\xi(\gamma_2\gamma_1, \gamma_3, \gamma_2)|^2 d\lambda^{r(\gamma_2)}(\gamma_3) d\lambda^{s(\gamma_2)}(\gamma_1) d\lambda'_x(\gamma_2) d\mu(x)$$
$$= \iiiint |\xi(\gamma_1, \gamma_3, \gamma_2)|^2 d\lambda^{r(\gamma_2)}(\gamma_3) d\lambda^{r(\gamma_2)}(\gamma_1) d\lambda^x(\gamma_2) d\mu(x)$$
$$= \|\xi\|^2.$$

Thus U_0 is actually an isometry. The adjoint is given by

$$\{U_0^*\eta\}(\gamma_1, \gamma_2, \gamma_3) = \eta(\gamma_3^{-1}\gamma_1, \gamma_3, \gamma_3^{-1}\gamma_2)$$

for any $\eta \in \mathcal{H}_2$ and $(\gamma_1, \gamma_2, \gamma_3)$ with $r(\gamma_1) = r(\gamma_2) = r(\gamma_3)$. Hence U_0 is indeed a unitary.

For any $\xi \in \mathcal{H}$, $\eta \in \mathcal{H}_1$ and $(\gamma_1, \gamma_2, \gamma_3)$ with $r(\gamma_1) = r(\gamma_2) = r(\gamma_3)$, we compute

$$\{(1 \otimes U_0^*)\Phi(A_{f,q})(1 \otimes U_0)(\xi \otimes \eta)\}(\gamma_1, \gamma_2, \gamma_3)$$
$$= \{\Phi(A_{f,q})(\xi \otimes U_0\eta)\}(\gamma_3^{-1}\gamma_1, \gamma_3, \gamma_3^{-1}\gamma_2)$$
$$= \int f(\gamma_3)q(\gamma)\{\xi \otimes U_0\eta\}(\gamma^{-1}\gamma_3^{-1}\gamma_1, \gamma_3\gamma, \gamma^{-1}\gamma_3^{-1}\gamma_2) d\lambda^{s(\gamma_3)}(\gamma)$$
$$= \int f(\gamma_3)q(\gamma)\eta(\gamma_1, \gamma_2, \gamma_3\gamma)\xi d\lambda^{s(\gamma_3)}(\gamma)$$
$$= \int f(\gamma_3)\check{q}(\gamma^{-1})\eta(\gamma_1, \gamma_2, \gamma_3\gamma)\xi d\lambda^{s(\gamma_3)}(\gamma)$$
$$= \{(1_{\mathcal{K}} \otimes_{\mathcal{Z}} 1_{L^2(\mathcal{G},\nu)} \otimes_{\mathcal{Z}} M(f)\rho(\check{q}))(\xi \otimes \eta)\}(\gamma_1, \gamma_2, \gamma_3),$$

where ρ is the right multiplication associated with the left Hilbert algebra $\mathcal{A}_I$. This computation together with (7.4) yields

$$(1 \otimes U_0)^*\Phi(A_{f,q})(1 \otimes U_0) = 1_{\mathcal{K}} \otimes_{\mathcal{Z}} 1_{L^2(\mathcal{G},\nu)} \otimes_{\mathcal{Z}} M(f)\rho(\check{q})$$
$$= (\delta *_{\mathcal{Z}} \iota)(1_{\mathcal{K}} \otimes_{\mathcal{Z}} M(f)\rho(\check{q})). \quad (7.5)$$

Also we calculate

$$
\begin{aligned}
&\{(1\otimes U_0)^*\Phi(A_{B,q})(1\otimes U_0)(\xi\otimes\eta)\}(\gamma_1,\gamma_2,\gamma_3)\\
&=\{\Phi(A_{B,q})(\xi\otimes U_0\eta)\}(\gamma_3^{-1}\gamma_1,\gamma_3,\gamma_3^{-1}\gamma_2)\\
&=\int q(\gamma)\Theta(B)_{r(\gamma)}(\xi\otimes U_0\eta)(\gamma^{-1}\gamma_3^{-1}\gamma_1,\gamma_3\gamma,\gamma^{-1}\gamma_3^{-1}\gamma_2)d\lambda^{s(\gamma_3)}(\gamma)\\
&=\iint q(\gamma)\{B(\gamma_0)\xi\}\{U_0\eta\}(\gamma^{-1}\gamma_3^{-1}\gamma_1,\gamma_0^{-1}\gamma_3\gamma,\gamma^{-1}\gamma_3^{-1}\gamma_2)\\
&\qquad\qquad d\lambda^{r(\gamma_3)}(\gamma_0)d\lambda^{s(\gamma_3)}(\gamma)\\
&=\iint q(\gamma)\{B(\gamma_0)\xi\}\eta(\gamma_0^{-1}\gamma_1,\gamma_0^{-1}\gamma_2,\gamma_0^{-1}\gamma_3\gamma)d\lambda^{r(\gamma_3)}(\gamma_0)d\lambda^{s(\gamma_3)}(\gamma)\\
&=\{\Phi_0''(B_\lambda\otimes\lambda\otimes\lambda)\big(1_{\mathcal{K}}\otimes_{\mathcal{Z}}1_{L^2(\mathcal{G},\nu)}\otimes_{\mathcal{Z}}\rho(\check{q})\big)(\xi\otimes\eta)\}(\gamma_1,\gamma_2,\gamma_3).
\end{aligned}
$$

From this calculation together with (7.4), it follows that

$$
\begin{aligned}
(1\otimes U_0)^*\Phi(A_{B,q})(1\otimes U_0)&=\Phi_0''(B_\lambda\otimes\lambda\otimes\lambda)\big(1_{\mathcal{K}}\otimes_{\mathcal{Z}}1_{L^2(\mathcal{G},\nu)}\otimes_{\mathcal{Z}}\rho(\check{q})\big)\\
&=(\delta*_{\mathcal{Z}}\iota)\big(\delta\big(\Phi_0(B_\lambda)\big)(1_{\mathcal{K}}\otimes_{\mathcal{Z}}\rho(\check{q}))\big). \qquad (7.6)
\end{aligned}
$$

Thus we obtain

Proposition 7.7. *Let* $\pi=\mathrm{Ad}(1\otimes U_0)\circ(\delta*_{\mathcal{Z}}\iota)$. *Then* π *gives a* $*$*-isomorphism from* $\mathcal{N}*_{\mathcal{Z}}\mathcal{L}({}_{\mathcal{Z}}L^2(\mathcal{G},\nu))$ *onto the double crossed product algebra* $(\mathcal{N}\times_\delta\mathcal{G})\times_{\hat{\delta}}\mathcal{G}$.

Proof. This immediately follows from the combination of Proposition 7.3, (7.5) and (7.6). Q.E.D.

On $(\mathcal{N}\times_\delta\mathcal{G})\times_{\hat{\delta}}\mathcal{G}$, we have the bidual coaction $\tilde{\delta}$. Hence if we put $\bar{\delta}=(\pi^{-1}*_{\mathcal{Z}}\iota)\circ\tilde{\delta}\circ\pi$, then, by Corollary 1.3, $\bar{\delta}$ turns out to be a coaction of $\mathcal{G}$ on $\mathcal{N}*_{\mathcal{Z}}\mathcal{L}({}_{\mathcal{Z}}L^2(\mathcal{G},\nu))$. Then it is clear that we have

Theorem 7.8. **(Duality for integrable coactions)** *Let* $(\mathcal{G},\ \mathcal{N},\ \delta,\ \mathcal{K})$ *be an integrable coaction. Then the bidual coaction* $(\mathcal{G},\ (\mathcal{N}\times_\delta\mathcal{G})\times_{\hat{\delta}}\mathcal{G},\ \tilde{\delta})$ *is conjugate to a coaction* $(\mathcal{G},\ \mathcal{N} *_Z \mathcal{L}({}_Z L^2(\mathcal{G},\nu)),\ \bar{\delta})$.

Remark 7.9. In Definition 7.1 of integrability of a coaction, we may replace the condition there by a weaker one that $\{\mathcal{H}(x)\}_{x\in X}$ is constant only on each equivalence class on X, that is, $\mathcal{H}(x) = \mathcal{H}(y)$ only if $x \sim y$. Even if we relax the condition in this way, the above Theorem of duality for coactions still holds. Its proof can be obtained by a minor modification of our argument in this section.

§ 8. Examples of actions and coactions of measured groupoids on von Neumann algebras

In this section, we shall discuss several examples of actions and coactions of measured groupoids. Then we will apply the duality theorem to them.

Example 1. Let G be a (second countable) locally compact group. Suppose that we are given an action α of G on a (separable) von Neumann algebra $\mathcal{M}$. We choose a measure theoretic spectrum (X, μ) of the center $\mathcal{Z}$ of $\mathcal{M}$. We may assume that X is a separable compact space and that μ is a Borel measure on X quasi-invariant under the induced action of G on X. We set $\mathcal{G} = X \times G$. We suppose that G acts on X from the right. Then $\mathcal{G}$ becomes a (second countable) locally compact groupoid under the product topology with its multiplication and inverse defined as follows: $((x, g), (y, h)) \in \mathcal{G}^{(2)}$ if $y = xg$ and $(x, g)(xg, h) = (x, gh)$; the inverse is given by $(x, g)^{-1} = (xg, g^{-1})$. Since $s(x, g) = (xg, e)$ and $r(x, g) = (x, e)$, we may identify $\mathcal{G}^{(0)}$ with X. We define a transeverse function $\{\lambda^x\}_{x\in X}$ by $\lambda^x = \delta_x \times ds$, where ds indicates a left Haar measure of G, and δ_x is the point mass measure at x. Then we can turn $\mathcal{G}$ into a measured groupoid in an obvious manner. The measure ν in our notation corresponds to the product measure $\mu \times ds$.

Let $\{\mathcal{M}, \mathcal{H}_{\mathcal{M}}, J_{\mathcal{M}}, \mathcal{P}_{\mathcal{M}}\}$ be a standard form of the von Neumann algebra $\mathcal{M}$. We will simply write $\mathcal{H}$ for $\mathcal{H}_{\mathcal{M}}$ if there is no danger of confusion. We consider the central decomposition of $\mathcal{M}$ with respect to the spectrum (X, μ):

$$\mathcal{H} = \int_X^{\oplus} \mathcal{H}(x) d\mu(x), \qquad \mathcal{M} = \int_X^{\oplus} \mathcal{M}(x) d\mu(x).$$

According to this decomposition, we obtain a family $\{\alpha_{x,g}\}_{(x,g)\in\mathcal{G}}$ of $*$-isomorphisms from $\mathcal{M}(xg)=\mathcal{M}(s(x,g))$ onto $\mathcal{M}(x)=\mathcal{M}(r(x,g))$ satisfying

$$\alpha_{x,g}\,\alpha_{xg,h}=\alpha_{x,gh}, \qquad \alpha_{x,g}^{-1}=\alpha_{xg,g^{-1}},$$

for any $s, t\in G$ and $x\in X$. Since $\mathcal{M}(x)$ acts on $\mathcal{H}(x)$ standardly, we get a family $\{U(x,g)\}_{(x,g)\in\mathcal{G}}$ of unitary operators from $\mathcal{H}(xg)=\mathcal{H}(s(x,g))$ onto $\mathcal{H}(x)=\mathcal{H}(r(x,g))$ satisfying $U(x,g)\,U(xg,h)=U(x,gh)$ for any $s, t\in G$ and μ a.e.$x\in X$. Deleting a set of measure zero from X, we may assume that this identity holds for any $x\in X$. Then it is clear that $\{U(x,g)\}$ is a representation of the groupoid $\mathcal{G}$. Therefore, the system $(\mathcal{G},\ \{\mathcal{M}(x),\mathcal{H}(x)\},\ \{\alpha_{x,g}=\mathrm{Ad}\,U(x,g)\})$ is an action of $\mathcal{G}$. We note that $L^2(\mathcal{G},\nu)$ is nothing but $L^2(X,\mu)\otimes L^2(G,ds)$. Hence, for example, $\mathcal{H}\otimes_{\mu\mathcal{Z}}L^2(\mathcal{G},\nu)$ is (isomorphic to) the Hilbert space $\mathcal{H}\otimes L^2(G,ds)$. It can be shown without difficulty that the crossed product algebra $\mathcal{M}\times_\alpha\mathcal{G}$ is exactly the same as the usual crossed product algebra $\mathcal{M}\times_\alpha G$ obtained from the action α of G on $\mathcal{M}$. Moreover, the dual coaction of the groupoid $\mathcal{G}$ on the crossed product algebra is nothing other than the coaction of G on the crossed product algebra $\mathcal{M}\times_\alpha G$. Finally, we should remark that our duality theorem for actions is, in this case, Takesaki duality for actions of locally compact groups on von Neumann algebras.

Example 2. As usual, we let $\mathcal{G}$ be a measured groupoid with a system (λ,Λ,δ). Let $L^2(X,\mu)=\int_X^\oplus \mathbf{C}_x d\mu(x)$ be the direct integral decomposition of $L^2(X,\mu)$ with respect to $\mathcal{Z}=L^\infty(X,\mu)$, where $\mathbf{C}_x=\mathbf{C}$ is a one dimensional Hilbert space. Through this decomposition, we have $\mathcal{Z}=\int_X^\oplus \mathbf{C}(x)d\mu(x)$, where $\mathbf{C}(x)=\mathbf{C}$ acts on $\mathbf{C}_x$ by scalar multiplication. For every $\gamma\in\mathcal{G}$, we set $\iota_\gamma:\mathbf{C}(s(\gamma))\longrightarrow\mathbf{C}(r(\gamma))$ to be the identity morphism. One can easily see that the system $(\mathcal{G},\ \{\mathbf{C}(x), \mathbf{C}_x\},\ \{\iota_\gamma = \mathrm{Ad}\, 1\})$ is an action of $\mathcal{G}$. It is readily shown that the crossed product algebra $L^\infty(X,\mu)\times_\alpha \mathcal{G}$ is the groupoid von Neumann algebra $\mathcal{R}(\mathcal{G})$. The dual coaction $\hat{\iota}$ is the coproduct Γ of $\mathcal{R}(\mathcal{G})$.

Example 3. By the consequence of § 2, the system $(\mathcal{G},\ \mathcal{R}(\mathcal{G}),\ \Gamma,\ L^2(\mathcal{G},\nu))$ forms a coaction of $\mathcal{G}$. From the duality theorem for actions of $\mathcal{G}$, together with Example 2, it follows that the crossed product algebra $\mathcal{R}(\mathcal{G})\times_\Gamma \mathcal{G}$ associated with this coaction is exactly equal to

$$\begin{aligned}\{\, L^\infty(X,\mu)\times_\iota \mathcal{G}\,\}\times_\iota \mathcal{G} &\cong L^\infty(X,\mu)\otimes_Z \mathcal{Z}'_S \\ &\cong \mathcal{Z}'_S.\end{aligned}$$

Moreover, the dual action $\hat{\Gamma}_\gamma$ is the same as $\mathrm{Ad}\,\rho(\gamma)$. Consequently, $(\mathcal{G},\ \{\mathcal{L}(\, L^2(\mathcal{G}_x, \lambda'_x)\,),\ L^2(\mathcal{G}_x,\lambda'_x)\},\ \{\mathrm{Ad}\,\rho(\gamma)\})$ is the dual system of the coaction $(\mathcal{G},\ \mathcal{R}(\mathcal{G}),\ \Gamma,\ L^2(\mathcal{G},\nu))$.

Example 4. In the previous example, we saw that $\mathcal{Z}'_S$ is obtained as the crossed product algebra associated with a <u>coaction</u> of $\mathcal{G}$. In the following, we shall show that this algebra $\mathcal{Z}'_S$ can be obtained as the crossed product algebra derived from an <u>action</u> of $\mathcal{G}$. For this, we consider a system $(\mathcal{G},\ \{L^\infty(\mathcal{G}^x,\lambda^x),\, L^2(\mathcal{G}^x,\lambda^x)\},\ \{\lambda_\gamma = \mathrm{Ad}\,\lambda(\gamma)\})$, which is clearly an action of $\mathcal{G}$. Let $\mathcal{A} = L^\infty(\mathcal{G},\nu)$. Then, with the notation in § 4, we have $\mathcal{A}(\gamma) = L^\infty(\mathcal{G}^{r(\gamma)},\lambda^{r(\gamma)})\lambda(\gamma)\otimes\lambda(\gamma),\quad (\gamma\in\mathcal{G})$. It then follows from an algebraic calculation that the crossed product algebra $\mathcal{A}\times_\lambda \mathcal{G}$ is generated by $L^\infty(\mathcal{G},\nu)\otimes_Z\mathbf{C}$ and $\Gamma(\,\mathcal{R}(\mathcal{G})\,)$ on ${}_ZL^2(\mathcal{G},\nu)\otimes_\mu {}_ZL^2(\mathcal{G},\nu)$. Since $\{\mathrm{Ad}\,\sigma_{2,\mu}\}\circ\Gamma = \Gamma$, we have

$$\begin{aligned}\sigma_{2,\mu}\{\,\mathcal{A}\times_\lambda\mathcal{G}\,\}\sigma_{2,\mu} &= \sigma_{2,\mu}\{\,L^\infty(\mathcal{G},\nu)\otimes_Z\mathbf{C}\ \vee\ \Gamma(\,\mathcal{R}(\mathcal{G}))\,\}\sigma_{2,\mu} \\ &= \mathbf{C}\otimes_Z L^\infty(\mathcal{G},\nu)\ \vee\ \Gamma(\,\mathcal{R}(\mathcal{G})\,) \\ &= \mathcal{R}(\mathcal{G})\times_\Gamma\mathcal{G}.\end{aligned}$$

Hence, by the previous example, $\mathcal{A} \times_\lambda \mathcal{G}$ is isomorphic to $\mathcal{Z}'_S$. Therefore, we have a coaction on $\mathcal{Z}'_S$ derived from the dual coaction $\hat{\lambda}$ on $\mathcal{A} \times_\lambda \mathcal{G}$. We denote it by $\hat{\lambda}$ again. Note that $\mathcal{R}(\mathcal{G})$ is contained in $\mathcal{Z}'_S$. It can be verified that the restriction of $\hat{\lambda}$ to $\mathcal{R}(\mathcal{G})$ is nothing but the coproduct Γ of $\mathcal{R}(\mathcal{G})$.

Example 5. In [T4], Takesaki gave an example of an action of an ergodic hyperfinite measured equivalence relation (X, μ, R) with countable orbits, although he considered it in a context different from ours. It consists of a measurable field $\{\mathcal{M}(x), \mathcal{H}(x)\}_{x\in X}$ of von Neumann algebras and a family $\{\alpha_{(y,x)}\}_{(y,x)\in R}$ of $*$-isomorphisms from $\mathcal{M}(x)$ onto $\mathcal{M}(y)$ such that (i) all $\mathcal{M}(x)$ are factors of type III_0, (ii) $\mathcal{M}(y) = \mathcal{M}(x)$ if $(y, x) \in R$ (iii) $\alpha_{(y,x)} = \iota$ for any $(y, x) \in R$. In this case, the family $\{\alpha_{(y,x)}\}_{(y,x)\in R}$ gives rise to a centrally ergodic automorphism θ on the direct integral $\mathcal{M} = \int_X^\oplus \mathcal{M}(x) d\mu(x)$. Then we obtain a factor $\mathcal{P}$ of type III by taking the usual crossed product of $\mathcal{M}$ by θ:

$$\mathcal{P} = \mathcal{M} \times_\theta \mathbf{Z},$$

where $\mathbf{Z}$ indicates the ring of integers. It can be shown without much difficulty that the algebra $\mathcal{P}$ is captured by the crossed product construction from the groupoid action of (X, μ, R) in our sense. Namely, we have $\mathcal{P} = \mathcal{M} \times_\alpha R$. Moreover, the crossed product algebra $\mathcal{P} \times_{\hat{\theta}} \mathbf{T}$ ($\hat{\theta}$ is the dual action of θ) exactly coincides with the crossed product of $\mathcal{P}$ by the dual coaction $\hat{\alpha}$ of (X, μ, R) in our sense. Hence we have $\mathcal{P} \times_{\hat{\theta}} \mathbf{T} = \mathcal{P} \times_{\hat{\alpha}} R = (\mathcal{M} \times_\alpha R) \times_{\hat{\alpha}} R$. Here $\mathbf{T} = \hat{\mathbf{Z}}$ is the one dimensional torus.

Example 6. Let $\mathcal{G}$ be measured groupoid as usual. We assume here for the sake of our argument that $\mathcal{G}$ is a separable locally compact topological groupoid. According to

[Ma], the triple $(\Omega, \mathcal{G}, \alpha)$ is called a locally compact measured transformation groupoid if Ω is a locally compact space and the mapping $\alpha: \mathcal{G} \times \Omega \longrightarrow \Omega$ sending (γ, ω) to $\alpha_\gamma(\omega)$ is a continuous action of $\mathcal{G}$ on Ω, and if Ω is equipped with a Borel measure m quasi-invariant under the action of $\mathcal{G}$. We can turn $\tilde{\mathcal{G}} = \Omega \times \mathcal{G}$ into a measured groupoid in the following manner: a pair $((\omega_1, \gamma_1), (\omega_2, \gamma_2))$ of elements in $\tilde{\mathcal{G}} \times \tilde{\mathcal{G}}$ is multiplicative if $\omega_2 = \alpha_{\gamma_1^{-1}}(\omega_1)$ and $(\gamma_1, \gamma_2) \in \mathcal{G}^{(2)}$; the multiplication is then defined as $(\omega_1, \gamma_1)(\alpha_{\gamma_1^{-1}}(\omega_1), \gamma_2) = (\omega_1, \gamma_1\gamma_2)$; the inverse of (ω, γ) is $(\omega, \gamma)^{-1} = (\alpha_{\gamma^{-1}}(\omega), \gamma^{-1})$. It is immediate to check that $s(\omega, \gamma) = (\alpha_{\gamma^{-1}}(\omega), s(\gamma))$ and $r(\omega, \gamma) = (\omega, r(\gamma))$. Thus the unit space $\tilde{\mathcal{G}}^{(0)}$ may be naturally identified with $\Omega \times X$. The transverse function $\{\tilde{\lambda}^{(\omega, x)}\}_{(\omega, x) \in \Omega \times X}$ for this groupoid is obtained from $\{\lambda^x\}_{x \in X}$ as follows:

$$d\tilde{\lambda}^{(\omega, x)}(\omega, \gamma) = d\lambda^x(\gamma).$$

It is easy to see that the above transverse function is faithful and proper. We set

$$\tilde{\delta}(\omega, \gamma) = \delta(\gamma) \frac{dm}{dm \circ \alpha_{\gamma^{-1}}}(\omega),$$

where $m \circ \alpha_\gamma(E) = m(\alpha_{\gamma^{-1}}(E))$. Then it can be verified that we have $\tilde{\delta}((m \times \mu) \circ \tilde{\lambda})\check{} = (m \times \mu) \circ \tilde{\lambda}$. Hence, by virtue of Theorem 3 in [C3], there exists a transverse measure $\tilde{\Lambda}$ with module $\tilde{\delta}$ for the groupoid $\tilde{\mathcal{G}}$ such that $\tilde{\Lambda}_{\tilde{\lambda}} = m \times \mu$. Thus we get a measured groupoid $\tilde{\mathcal{G}}$ with a Haar measure $(\tilde{\lambda}, \tilde{\Lambda}, \tilde{\delta})$ (see also Lemma 5.4 of [Ma]).

Let $f \in L^\infty(\Omega, m)$ and $\gamma \in \mathcal{G}$. We define a $*$-automorphism, denoted again by α_γ, of $L^\infty(\Omega, m)$ by

$$\alpha_\gamma(f)(\omega) = f(\alpha_{\gamma^{-1}}(\omega)).$$

One can easily check that

$$\alpha_{\gamma_1} \circ \alpha_{\gamma_2} = \alpha_{\gamma_1\gamma_2}, \qquad \alpha_\gamma^{-1} = \alpha_{\gamma^{-1}}.$$

For each $\gamma \in \mathcal{G}$, we consider a unitary $u(\gamma)$ on $L^2(\Omega, m)$ given by

$$\{u(\gamma)\xi\}(\omega) = \sqrt{\frac{dm \circ \alpha_\gamma}{dm}(\omega)}\, \xi(\alpha_{\gamma^{-1}}(\omega)), \qquad (\, \xi \in L^2(\Omega, m)\,).$$

From the cocycle identity, it follows that this u is a representation of $\mathcal{G}$ on $L^2(\Omega, m)$. Furthermore, it is readily verified that this representation implements the automorphisms α_γ:

$$\alpha_\gamma = \operatorname{Ad} u(\gamma).$$

Set $\mathcal{M}(x) = L^\infty(\Omega, m)$ and $\mathcal{H}(x) = L^2(\Omega, m)$ for all $x \in X$. Then the system $(\mathcal{G}, \{\mathcal{M}(x), \mathcal{H}(x)\}, \{\alpha_\gamma = \operatorname{Ad} u(\gamma)\})$ is clearly an action of $\mathcal{G}$. By Example 2, we have that $\mathcal{R}(\tilde{\mathcal{G}}) = L^\infty(\Omega \times X,\, m \times \mu) \times_\lambda \tilde{\mathcal{G}}$. At the same time, it can be shown without difficulty that the crossed product algebra of the action obtained above coincides with $\mathcal{R}(\tilde{\mathcal{G}})$. Namely, we have

$$\mathcal{R}(\tilde{\mathcal{G}}) = L^\infty(\Omega, m) \times_\alpha \mathcal{G}.$$

Remark Proposition 5.5 in [**Ma**] for the similar result.

In the following, we shall give an example of the above notion of a locally compact measured transformation groupoid. The example is due to A. Connes [**C2**]. See also [**Ma**], [**MS**], [**Se**].

Let G be the simple Lie group $SL(2, \mathbf{R})$. We take a discrete cocompact subgroup Γ of $SL(2, \mathbf{R})$. Then we look at the smooth manifold $M = SL(2, \mathbf{R})/\Gamma$. Set H_1 to be the subgroup of $SL(2, \mathbf{R})$ consisting of matrices A of the form:

$$A = \begin{pmatrix} 1 & 0 \\ t & 1 \end{pmatrix},$$

for $t \in \mathbf{R}$. We also set H_2 to be the set of all matrices A in $SL(2, \mathbf{R})$ of the form:

$$A = \begin{pmatrix} e^t & 0 \\ 0 & e^{-t} \end{pmatrix},$$

for $t \in \mathbf{R}$. Both H_1 and H_2 are one-parameter subgroups of $SL(2, \mathbf{R})$. Hence they generate flows on the manifold M via left translation. The flow given by the subgroup H_1 is called the horocycle flow, which can be viewed as the horocylce flow on the unit tangent bundle of a surface of constant negative sectional curvature. The other flow generated by the subgroup H_2 is called the geodesic flow. This flow is well-known to be an Anosov flow. Hence it gives rise to the Anosov foliation $\mathcal{F}_0$ on M. Let

$$N = \left\{ \begin{pmatrix} e^t & 0 \\ a & e^{-t} \end{pmatrix} : t, a \in \mathbf{R} \right\}.$$

Then the Anosov foliation is given by the orbits of the action of N on M. It is shown in [**B**] (see also [**Se**]) that the locally compact measured groupoid $\mathcal{G}$ constructed as the graph of this foliation $\mathcal{F}_0$ is hyperfinite and of type III_1. Let $(K, \mathbf{R}, F_t)$ be an ergodic (measure-preserving) smooth flow on a compact manifold K with a probability measure. To this system, we associate a new foliation $\mathcal{F}$ on $K \times M$ derived from the action of the group N, where

$$\begin{pmatrix} 1 & 0 \\ a & 1 \end{pmatrix}$$

acts trivially on K and by the horocycle action on M, and

$$\begin{pmatrix} e^t & 0 \\ 0 & e^{-t} \end{pmatrix}$$

acts by F_t on K and by the geodesic flow on M. It is noticed in [**Ma**] that the graph $\tilde{\mathcal{G}}$ of the resulting foliation $\mathcal{F}$ is equal to the locally compact measured transformation groupoid $K \times_\alpha \mathcal{G}$, where $\alpha(\omega, (A, x)) = F_{\pi(A)}(\omega)$. Here $\pi : N \mapsto \mathbf{R}$ is defined by

$$\begin{pmatrix} e^t & 0 \\ a & e^{-t} \end{pmatrix} \in N \longmapsto t \in \mathbf{R}.$$

By the fact noted in the previous paragraph, the associated groupoid von Neumann algebra $\mathcal{R}(\tilde{\mathcal{G}})$ of $\tilde{\mathcal{G}}$ is isomorphic to the crossed product algebra $L^\infty(K) \times_\alpha \mathcal{G}$. This von Neumann algebra is a factor of type III with the smooth flow of weight [**CT**] isomorphic to the system $(K,\ \mathbf{R},\ F_t)$. If we take K to be the circle of length L and F_t to be rotations with speed 1, then the von Neumann algebra is of type III_λ, where $\lambda = \exp(-L)$. If $\dim K > 1$, then the algebra is of type III_0.

Example 7. Let $(\mathcal{G},\ \lambda,\ \Lambda,\ \delta)$ be a measured groupoid. We choose a symmetric probability measure τ from the measure class $[\nu]$. Here, by "symmetric", we mean that $\tau^{-1} = \tau$, where τ^{-1} denotes the image of the measure τ by the inverse map $\gamma \mapsto \gamma^{-1}$ of $\mathcal{G}$. Let $\mathcal{G}'$ be the principal groupoid derived from $\mathcal{G}$. In other words, $\mathcal{G}'$ is the graph of the equivalence relation on X. We put $\tau' = P_*(\tau)$, where P is the Borel map from $\mathcal{G}$ onto the groupoid $\mathcal{G}'$ defined by $P(\gamma) = (r(\gamma),\ s(\gamma))$. Also set $\tilde{\tau} = r_*(\tau)$ as usual. Let $\tau' = \int \tau'_x d\tilde{\tau}(x)$ be a decomposition of τ' relative to the projection p_1 onto the first coordinate. By Theorem 3.9 of [**Ha1**], there are a conull Borel subset X_0 of X and a Borel function $p : \mathcal{G}'_{X_0} \longrightarrow \mathbf{R}_+$ such that, for all Borel $f \geq 0$ on $\mathcal{G}'_{X_0}$ and $(x_1,\ y_1) \in \mathcal{G}'_{X_0}$,

$$\int f((x_1,\ y_1)(x,\ y))p(x,\ y)d\tau'_{y_1}(x,\ y) = \int f(x_1,\ y)p(x,\ y)d\tau'_{x_1}(x,\ y),$$

and the function $(x, y) \in \mathcal{G}'_{X_0} \mapsto \delta'(x, y) = p(x, y)/p(y, x) \in (0, \infty)$ is a strict homomorphism of $\mathcal{G}'_{X_0}$. Hence if we set $d\nu'_x = p d\tau'_x$ and $\tau'_1 = \int \nu'_x d\tilde{\tau}(x)$, then the pair $(\tau'_1,\ \tilde{\tau})$ is a Haar measure in the sense of P. Hahn [**Ha1**]. For the sake of simplicity of our argument, in the remainder of this example, we will replace $\mathcal{G}'$ by its inessential reduction $\mathcal{G}'_{X_0}$. So we have a measured groupoid $\mathcal{G}'$ with a system $(\{\nu'_x\}_{x\in X}, \tilde{\tau}, \delta')$.

Now, since $\nu = \int \lambda^x d\mu(x) = \int \lambda^x d\mu/d\tilde{\tau} d\tilde{\tau}(x)$, $L^2(\mathcal{G},\nu)$ can be decomposed relative to $\mathcal{Z}$: ${}_{\mathcal{Z}}L^2(\mathcal{G},\nu) = \int_X^{\oplus} L^2(\mathcal{G}^x, d\mu/d\tilde{\tau}(x)\ \lambda^x)d\tilde{\tau}(x)$. We have a canonical (left) action of $\mathcal{Z}$ on $L^2(\mathcal{G}',\tau'_1)$ by $h \in \mathcal{Z} \mapsto M(h\circ p_1)$. We write ${}_{\mathcal{Z}}L^2(\mathcal{G}',\tau'_1)$ for $L^2(\mathcal{G}',\tau'_1)$ if this $\mathcal{Z}$-action is especially considered on it. Then we have a decomposition ${}_{\mathcal{Z}}L^2(\mathcal{G}',\tau'_1) = \int_X^{\oplus} L^2(\mathcal{G}'^x,\nu'_x)\ d\tilde{\tau}(x)$. Hence we have ${}_{\mathcal{Z}}L^2(\mathcal{G},\nu)\otimes_{\tilde{\tau}}\ {}_{\mathcal{Z}}L^2(\mathcal{G}',\tau'_1) = \int_X^{\oplus} L^2(\mathcal{G}^x, d\mu/d\tilde{\tau}(x)\ \lambda^x) \otimes L^2(\mathcal{G}'^x,\nu'_x)\ d\tilde{\tau}(x)$ and $L^2(\mathcal{G},\nu)_{\mathcal{Z}} \otimes_{\tilde{\tau}}\ {}_{\mathcal{Z}}L^2(\mathcal{G}',\tau'_1) = \int_X^{\oplus} L^2(\mathcal{G}_x, d\mu/d\tilde{\tau}(x)\ \lambda'_x) \otimes L^2(\mathcal{G}'^x,\nu'_x)\ d\tilde{\tau}(x)$.
We consider two Borel subsets S_1, S_2 of $\mathcal{G}\times\mathcal{G}'$ given by

$$S_1 = \{\,(\gamma,x,y)\in \mathcal{G}\times\mathcal{G}' : r(\gamma) = x\,\},$$

$$S_2 = \{\,(\gamma,x,y)\in \mathcal{G}\times\mathcal{G}' : s(\gamma) = x\,\}.$$

We put measures $d\omega_1$, $d\omega_2$ on S_1, S_2 respectively by the following equations:

$$\int f_1(\gamma,x,y)d\omega_1(\gamma,x,y) = \iiint f_1(\gamma,u,v)\frac{d\mu}{d\tilde{\tau}}(x)d\lambda^x(\gamma)d\nu'_x(u,v)d\tilde{\tau}(x),$$

$$\int f_2(\gamma,x,y)d\omega_2(\gamma,x,y) = \iiint f_2(\gamma,u,v)\frac{d\mu}{d\tilde{\tau}}(x)d\lambda'_x(\gamma)d\nu'_x(u,v)d\tilde{\tau}(x),$$

for any $f_i \in \mathcal{F}(S_i)$ $(i = 1, 2)$.

Then we may identify the Hilbert spaces ${}_{\mathcal{Z}}L^2(\mathcal{G},\nu)\otimes_{\tilde{\tau}}\ {}_{\mathcal{Z}}L^2(\mathcal{G}',\tau'_1)$ and $L^2(\mathcal{G},\nu)_{\mathcal{Z}} \otimes_{\tilde{\tau}}\ {}_{\mathcal{Z}}L^2(\mathcal{G}',\tau'_1)$ with $L^2(S_1,d\omega_1)$ and $L^2(S_2,d\omega_2)$ respectively. There exists a unitary W_0 from $L^2(S_1,d\omega_1)$ onto $L^2(S_2,d\omega_2)$ defined by

$$\{W_0\xi\}(\gamma,x,y) = \xi(\gamma,r(\gamma),y).$$

In fact, we have

$$\int |\xi(\gamma,r(\gamma),y)|^2 d\omega_2(\gamma,x,y)$$
$$= \int\int\int |\xi(\gamma,r(\gamma),v)|^2\frac{d\mu}{d\tilde{\tau}}(x)\lambda'_x(\gamma)d\nu'_x(u,v)d\tilde{\tau}(x)$$

$$= \int\int\int |\xi(\gamma, r(\gamma), v)|^2 d\nu'_x(u,v)\lambda'_x(\gamma)d\mu(x)$$

$$= \int\int\int |\xi(\gamma, r(\gamma), v)|^2 d\nu'_x(u,v)d\lambda^x(\gamma)d\mu(x)$$

$$= \int\int\int |\xi(\gamma, r(\gamma), v)|^2 d\nu'_x(u,v)\frac{d\mu}{d\tilde{\tau}}(x)d\lambda^x(\gamma)d\tilde{\tau}(x)$$

$$= \int\int\int |\xi(\gamma, u, v)|^2 \frac{d\mu}{d\tilde{\tau}}(x)d\lambda^x(\gamma)d\nu'_x(u,v)d\tilde{\tau}(x)$$

$$= \int |\xi(\gamma, u, v)|^2 d\omega_1(\gamma, u, v) = \|\xi\|^2.$$

So W_0 is an isometry. It is easily checked that the adjoint is given by $\{W_0^*\eta\}(\gamma,x,y) = \eta(\gamma, s(\gamma), y)$ and is also an isometry. Thus W_0 is a unitary. On $L^2(S_2, d\omega_2) = L^2(\mathcal{G},\nu)_Z \otimes_{\tilde{\tau}} {}_Z L^2(\mathcal{G}', \tau_1')$, we may consider $a \otimes_Z 1$ for any $a \in \mathcal{R}(\mathcal{G})$. We define a map δ from $\mathcal{R}(\mathcal{G})$ into $\mathcal{L}(L^2(S_1, d\omega_1))$ by

$$\delta(a) = W_0^*(a \otimes_Z 1)W_0, \qquad (\, a \in \mathcal{R}(\mathcal{G})\,).$$

Since the couple $(\tau_1', \tilde{\tau})$ is a Haar measure for the groupoid $\mathcal{G}'$, we may consider the groupoid von Neumann algebra $\mathcal{R}(\mathcal{G}')$ on $L^2(\mathcal{G}', \tau_1')$ associated with that Haar measure. On $L^2(S_1, d\omega_1) = {}_Z L^2(\mathcal{G}, \nu) \otimes_{\tilde{\tau}} {}_Z L^2(\mathcal{G}', \tau_1')$, we can consider the fiber product $\mathcal{R}(\mathcal{G}) *_Z \mathcal{R}(\mathcal{G}')$. In what follows, we shall show that the image of the above $*$-isomorphism δ is contained in $\mathcal{R}(\mathcal{G}) *_Z \mathcal{R}(\mathcal{G}')$ and moreover, we will prove that the morphism δ gives a coaction of $\mathcal{R}(\mathcal{G}')$ on $\mathcal{R}(\mathcal{G})$.

Let $\xi \in L^2(S_1, d\omega_1)$ and $f \in \mathcal{A}_I$. Then

$$\begin{aligned}\{\delta(L(f))\xi\}(\gamma, x, y) &= \{W_0^*(L(f) \otimes_Z 1)W_0\xi\}(\gamma, x, y)\\ &= \{(L(f) \otimes_Z 1)W_0\xi\}(\gamma, s(\gamma), y)\\ &= \int f(\gamma_1)\{W_0\xi\}(\gamma_1{}^{-1}\gamma, s(\gamma), y)d\lambda^x(\gamma_1)\end{aligned}$$

$$= \int f(\gamma_1)\xi(\gamma_1{}^{-1}\gamma, s(\gamma_1), y) d\lambda^x(\gamma_1)$$

$$= \int f(\gamma_1)\xi(\gamma_1{}^{-1}\gamma, P(\gamma_1)^{-1}(x, y)) d\lambda^x(\gamma_1). \qquad (*)$$

We define a canonical (i.e. the regular) representation λ' of $\mathcal{G}'$ on $\{L^2(\mathcal{G}'^x, \nu'_x)\}_{x\in X}$ over (X, μ) by

$$\{\lambda'(y, x)\xi\}(y, z) = \xi(x, z), \qquad (\xi \in L^2(\mathcal{G}'^x, \nu'_x),\ (y, x) \in \mathcal{G}'^y\,).$$

If $g \in L^2(\mathcal{G}, \nu)$ is $\tilde{\tau}$-bounded vector and $\eta \in L^2(\mathcal{G}', \tau'_1)$, and if ξ in $(*)$ is of the form $\xi = g \otimes_{\tilde{\tau}} \eta$, then we obtain

$$\{\delta(L(f))\xi\}(\gamma, x, y) = \int f(\gamma_1)\{\lambda(\gamma_1)g\}(\gamma)\{\lambda'(P(\gamma_1))\eta\}(x, y) d\lambda^x(\gamma_1). \qquad (*)'$$

Now if $T_1 \in \mathcal{R}(\mathcal{G})'$ and $T_2 \in \mathcal{R}(\mathcal{G}')'$, then, along the direct integral decompositions of $_Z L^2(\mathcal{G}, \nu) = \int_X^{\oplus} L^2(\mathcal{G}^x, d\mu/d\tilde{\tau}(x)\lambda^x) d\tilde{\tau}(x)$ and $_Z L^2(\mathcal{G}', \nu'_1) = \int_X^{\oplus} L^2(\mathcal{G}'^x, \nu'_x)\, d\tilde{\tau}(x)$, T_1 and T_2 are decomposed into

$$T_1 = \int_X^{\oplus} T_1(x) d\tilde{\tau}(x), \qquad T_2 = \int_X^{\oplus} T_2(x) d\tilde{\tau}(x)$$

such that we have $T_1(r(\gamma))\, \lambda(\gamma) = \lambda(\gamma)\, T_1(s(\gamma))$, $(\gamma \in \mathcal{G})$ and $T_2(y)\, \lambda'(y, x) = \lambda'(y, x)\, T_2(x)$, $((y, x) \in \mathcal{G}')$. Then, from $(*)'$, it follows that $\delta(L(f))(T_1 \otimes_Z T_2) = (T_1 \otimes_Z T_2)\, \delta(L(f))$. Thus $\delta(L(f))$ belongs to the commutant of the von Neumann algebra $\mathcal{R}(\mathcal{G})' \otimes_Z \mathcal{R}(\mathcal{G}')'$ generated by such elements $T_1 \otimes_Z T_2$. Hence $\delta(L(f))$ lies in the fiber product $\mathcal{R}(\mathcal{G}) *_Z \mathcal{R}(\mathcal{G}')$. Consequently, we have $\delta(\mathcal{R}(\mathcal{G})) \subseteq \mathcal{R}(\mathcal{G}) *_Z \mathcal{R}(\mathcal{G}')$.

Our next aim is to show that δ satisfies the coassociativity with respect to the coproduct Γ' of $\mathcal{R}(\mathcal{G}')$: $(\delta *_Z \iota) \circ \delta = (\iota *_Z \Gamma') \circ \delta$.

Let $\eta \in {}_{\mathcal{Z}}L^2(\mathcal{G}'^x, \tau_1')$ be a $\tilde{\tau}$-bounded element. Note that, if $\eta = \{\eta_x\}_{x\in X}$ is a decomposition of η through the direct integral ${}_{\mathcal{Z}}L^2(\mathcal{G}', \tau_1') = \int_X^{\oplus} L^2(\mathcal{G}'^x, \nu_x')d\tilde{\tau}(x)$, then the function $x \in X \longmapsto \|\eta_x\|$ is essentially bounded. For such a η, we define an operator $S_\eta : L^2(\mathcal{G}, \nu) \longrightarrow L^2(S_1, d\omega_1)$ by

$$\{S_\eta g\}(\gamma, x, y) = \eta(s(\gamma), y)g(\gamma),$$

for any $g \in L^2(\mathcal{G}, \nu)$ and $(\gamma,\, x,\, y) \in S_1$. Since

$$\begin{aligned}
&\int\!\!\int\!\!\int |\eta(s(\gamma), v)|^2 |g(\gamma)|^2 \frac{d\mu}{d\tilde{\tau}}(x) d\lambda^x(\gamma) d\nu_x'(u, v) d\tilde{\tau}(x) \\
&= \int\!\!\int\!\!\int |\eta(s(\gamma), v)|^2 |g(\gamma)|^2 d\nu_x'(u, v) d\lambda^x(\gamma) d\mu(x) \\
&\le \Big(\operatorname*{ess.\,sup}_{x\in X} \int |\eta(u, v)|^2 d\nu_x'(u, v)\Big) \|g\|^2 < \infty,
\end{aligned}$$

the operator S_η is bounded. Moreover, if f is a Borel function in $\mathcal{A}_I$, then

$$\begin{aligned}
\{S_\eta L(f) g\}(\gamma, x, y) &= \{S_\eta f * g\}(\gamma, x, y) \\
&= \eta(s(\gamma), y) \int f(\gamma_1) g({\gamma_1}^{-1}\gamma) d\lambda^x(\gamma).
\end{aligned}$$

Thus we obtain

$$\begin{aligned}
\{\delta(L(f)) S_\eta g\}(\gamma, x, y) &= \int f(\gamma_1)\{S_\eta g\}({\gamma_1}^{-1}\gamma, s(\gamma_1), y) d\lambda^x(\gamma_1) \\
&= \int f(\gamma_1) \eta(s(\gamma), y) g({\gamma_1}^{-1}\gamma) d\lambda^x(\gamma_1) \\
&= \{S_\eta L(f) g\}(\gamma, x, y).
\end{aligned}$$

This implies that $\delta(a)S_\eta = S_\eta\delta(a)$ for all $a \in \mathcal{R}(\mathcal{G})$. We also note that the span of $\{\, S_\eta g : \eta$ is $\tilde{\tau}$-bounded, $g \in L^2(\mathcal{G}, \nu) \,\}$ is dense in $L^2(S_1, d\omega_1)$, because $\{S_\eta g\}(\gamma, x, y) = \eta(s(\gamma), y)g(\gamma) = \{W_0^*(g \otimes \eta)\}(\gamma, x, y)$.

Now we consider a representation Ω of $\mathcal{G}$ on $\{L^2(\mathcal{G}^x, d\mu/d\tilde{\tau}(x)\lambda^x) \otimes L^2(\mathcal{G}'^x, \nu_x') \otimes L^2(\mathcal{G}'^x, \nu_x')\,\}_{x\in X}$ over $(X, \tilde{\tau})$, where Ω is defined by $\Omega(\gamma) = \lambda(\gamma) \otimes \lambda'(P(\gamma)) \otimes \lambda'(P(\gamma))$,

($\gamma \in \mathcal{G}$). Since P is a homomorphism from $\mathcal{G}$ onto $\mathcal{G}'$, this really defines a representation. To any Borel function $f \in \mathcal{A}_I$, we associate an "integrated form" $\Omega(f)$ on ${}_Z L^2(\mathcal{G},\nu) \otimes_{\tilde{\tau}} {}_Z L^2(\mathcal{G}', \tau_1') \otimes_{\tilde{\tau}} {}_Z L^2(\mathcal{G}',\tau_1') = \int_X^{\oplus} L^2(\mathcal{G}^x, d\mu/d\tilde{\tau}\lambda^x) \otimes L^2(\mathcal{G}'^x, \nu_x') \otimes L^2(\mathcal{G}'^x, \nu_x')\, d\tilde{\tau}(x)$ as follows; first we identify the above Hilbert space with $L^2(\tilde{S}_1, d\tilde{\omega}_1)$, where

$$\tilde{S}_1 = \{ (\gamma, (x,y), (u,v)) \in \mathcal{G} \times \mathcal{G}' \times \mathcal{G}' : r(\gamma) = x = u \}$$

$$d\tilde{\omega}_1(\gamma, (x,y), (z,w)) = \frac{d\mu}{d\tilde{\tau}}(x) d\lambda^x(\gamma) d\nu_x'(x,y) d\nu_x'(z,w) d\tilde{\tau}(x).$$

The identification is given by sending $g \otimes_{\tilde{\tau}} \eta_1 \otimes_{\tilde{\tau}} \eta_2$ to a function given by $(g \otimes_{\tilde{\tau}} \eta_1 \otimes_{\tilde{\tau}} \eta_2)(\gamma, (x,y), (u,v)) = g(\gamma)\eta_1(x,y)\eta_2(u,v)$. Then $\Omega(f)$ is defined to be

$$\begin{aligned}
&\{\Omega(f)\xi\}(\gamma, (x,y), (u,v)) \\
&= \int f(\gamma_1)\xi(\gamma_1{}^{-1}\gamma, (s(\gamma_1), y), (s(\gamma_1), v)) d\lambda^x(\gamma_1) \\
&= \int f(\gamma_1)\xi(\gamma_1{}^{-1}\gamma, P(\gamma_1)^{-1}(x,y), P(\gamma_1)^{-1}(u,v)) d\lambda^x(\gamma_1).
\end{aligned}$$

Then this satisfies the following identity.

Lemma 8.1. *We have the identity:*

$$(\delta *_Z \iota)\big(\delta(L(f))\big) = \Omega(f).$$

Proof. Let η_1, η_2 be $\tilde{\tau}$-bounded elements in ${}_Z L^2(\mathcal{G}', \tau_1')$ and $g \in L^2(\mathcal{G}, \nu)$. Also let $T_2 \in \mathcal{R}(\mathcal{G}')'$ as before. Then

$$\begin{aligned}
&\{(S_\eta \otimes_Z T_2)\, \delta\big(L(f)\big)(g \otimes_{\tilde{\tau}} \eta_1)\}(\gamma, (x,y), (u,v)) \\
&= \int f(\gamma_1)\eta(s(\gamma), y) g(\gamma_1{}^{-1}\gamma)\{T_2(u)\lambda'(P(\gamma_1))\eta_1\}(u,v) d\lambda^x(\gamma_1) \\
&= \int f(\gamma_1)\eta(s(\gamma), y) g(\gamma_1{}^{-1}\gamma)\{\lambda'(P(\gamma_1))T_2(v)\eta_1\}(u,v) d\lambda^x(\gamma_1).
\end{aligned}$$

In the meantime, we have

$$\{\Omega(f)(S_\eta \otimes_Z T_2)(g \otimes_{\tilde\tau} \eta_1)\}(\gamma, (x,y), (u,v))$$

$$= \{\Omega(f)(S_\eta g \otimes_{\tilde\tau} T_2\eta_1)\}(\gamma, (x,y), (u,v))$$

$$= \int f(\gamma_1)\{S_\eta g \otimes_{\tilde\tau} T_2\eta_1\}(\gamma_1{}^{-1}\gamma, (s(\gamma_1), y), (s(\gamma_1), v))d\lambda^x(\gamma_1)$$

$$= \int f(\gamma_1)\eta(s(\gamma), y)g(\gamma_1{}^{-1}\gamma)\{T_2\eta_1\}(s(\gamma_1), v)d\lambda^x(\gamma_1)$$

$$= \int f(\gamma_1)\eta(s(\gamma), y)g(\gamma_1^{-1}\gamma)\{\lambda'(P(\gamma_1))T_2(v)\eta_1\}(u,v)d\lambda^x(\gamma_1).$$

From this, it follows that

$$(S_\eta \otimes_Z T_2)\,\delta\big(L(f)\big)(g \otimes_{\tilde\tau} \eta_1) \;=\; \Omega(f)(S_\eta \otimes_Z T_2)(g \otimes_{\tilde\tau} \eta_1).$$

But, by Proposition 1.2, we have

$$(S_\eta \otimes_Z T_2)\delta\big(L(f)\big)(g \otimes_{\tilde\tau} \eta) = (\delta *_Z \iota)\big(\delta\big(L(f)\big)\big)(S_\eta \otimes_Z T_2)(g \otimes_{\tilde\tau} \eta_1).$$

Hence we obtain

$$(\delta *_Z \iota)\big(\delta\big(L(f)\big)\big)(S_\eta \otimes_Z T_2)(g \otimes_{\tilde\tau} \eta_1) = \Omega(f)(S_\eta \otimes_Z T_2)(g \otimes_{\tilde\tau} \eta_1).$$

Since the span of $\{\, (S_\eta \otimes_Z T_2)(g \otimes_{\tilde\tau} \eta_1) : \ \eta, \eta_1$ are $\tilde\tau$-bounded, $g \in L^2(\mathcal{G}, \nu),\ T_2 \in \mathcal{R}(\mathcal{G}')' \,\}$ forms a dense subspace of ${}_Z L^2(\mathcal{G}, \nu) \otimes_{\tilde\tau} {}_Z L^2(\mathcal{G}', \tau_1') \otimes_{\tilde\tau} {}_Z L^2(\mathcal{G}', \tau_1')$, we conclude that

$$(\delta *_Z \iota)\big(\delta\big(L(f)\big)\big) = \Omega(f).$$

Thus we get the desired identity. Q.E.D.

Let η be as above. This time, we define an operator R_η from $L^2(\mathcal{G}, \tau_1')$ onto ${}_Z L^2(\mathcal{G}', \tau_1') \otimes_{\tilde\tau} {}_Z L^2(\mathcal{G},, \tau_1')$ by

$$\{R_\eta \zeta\}((x,y),(x,z)) = \eta(y,z)\zeta(x,y), \quad (\zeta \in L^2(\mathcal{G}', \tau_1')),$$

where we identified the Hilbert space ${}_{\mathcal{Z}}L^2(\mathcal{G}',{}_{\tau}{}'_1) \otimes_{\tilde{\tau}} {}_{\mathcal{Z}}L^2(\mathcal{G}',{}_{\tau}{}'_1)$ with the square integrable functions on $\mathcal{H}'^{(2)} = \{\, ((x,y),(x,z)) \in \mathcal{G}' \times \mathcal{G}' : x \sim y \sim z \,\}$ with a measure determined by $\int \nu'_x \otimes \nu'_x d\tilde{\tau}(x)$. Then, if Γ' denotes the coproduct of $\mathcal{R}(\mathcal{G}')$, then it can be shown, as in the case of Γ, that $\Gamma'(a)R_\eta = R_\eta a, \ \ (a \in \mathcal{R}(\mathcal{G}'))$.

Lemma 8.2. *We have*

$$(\iota *_{\mathcal{Z}} \Gamma')\big(\delta(L(f))\big) = \Omega(f).$$

Proof. Let η, η_1 be $\tilde{\tau}$-bounded elements in ${}_{\mathcal{Z}}L^2(\mathcal{G}',{}_{\tau}{}'_1)$. Suppose that $g \in L^2(\mathcal{G},\nu)$ and $T_1 \in \mathcal{R}(\mathcal{G})'$ are as before. Then we calculate

$$\begin{aligned}
&\{\Omega(f)(T_1 \otimes_{\mathcal{Z}} R_\eta)(g \otimes_{\tilde{\tau}} \eta_1)\}(\gamma,(x,y),(x,z)) \\
&= \{\Omega(f)(T_1 g \otimes_{\tilde{\tau}} R_\eta \eta_1)\}(\gamma,(x,y),(x,z)) \\
&= \int f(\gamma_1)\{T_1 g\}(\gamma_1{}^{-1}\gamma)\{R_\eta \eta_1\}((s(\gamma_1),y),(s(\gamma_1),z))d\lambda^x(\gamma_1) \\
&= \int f(\gamma_1)\{\lambda(\gamma_1)T_1(s(\gamma_1))g\}(\gamma)\eta(y,z)\eta_1(s(\gamma_1),y)d\lambda^x(\gamma_1) \\
&= \int f(\gamma_1)\{T_1(x)\lambda(\gamma_1)g\}(\gamma)\eta(y,z)\eta_1(s(\gamma_1),y)d\lambda^x(\gamma_1).
\end{aligned}$$

We also compute

$$\begin{aligned}
&\{(T_1 \otimes_{\mathcal{Z}} R_\eta)\delta\big(L(f)\big)(g \otimes_{\tilde{\tau}} \eta_1)\}(\gamma,(x,y),(x,z)) \\
&= \int f(\gamma_1)\{T_1(x)\lambda(\gamma_1)g\}(\gamma)\{R_\eta \eta_1\}((s(\gamma_1),y),(s(\gamma_1),z))d\lambda^x(\gamma_1) \\
&= \int f(\gamma_1)\{T_1(x)\lambda(\gamma_1)g\}(\gamma)\eta(y,z)\eta_1(s(\gamma_1),y)d\lambda^x(\gamma_1).
\end{aligned}$$

This shows that $\Omega(f)(T_1 \otimes_{\mathcal{Z}} R_\eta)(g \otimes_{\tilde{\tau}} \eta_1) = (T_1 \otimes_{\mathcal{Z}} R_\eta)\delta\big(L(f)\big)(g \otimes_{\tilde{\tau}} \eta_1)$. But, by virtue of Proposition 1.2, we have

$$(T_1 \otimes_{\mathcal{Z}} R_\eta)\delta\big(L(f)\big)(g \otimes_{\tilde{\tau}} \eta_1) = (\iota *_{\mathcal{Z}} \Gamma')\big(\delta\big(L(f)\big)\big)(T_1 \otimes_{\mathcal{Z}} R_\eta)(g \otimes_{\tilde{\tau}} \eta_1).$$

It follows that we have

$$\Omega(f)(T_1 \otimes_{\mathcal{Z}} R_\eta)(g \otimes_{\tilde{\tau}} \eta_1) = (\iota *_{\mathcal{Z}} \Gamma')\big(\delta\big(L(f)\big)\big)(T_1 \otimes_{\mathcal{Z}} R_\eta)(g \otimes_{\tilde{\tau}} \eta_1).$$

Since the set $\{ (T_1 \otimes_{\mathcal{Z}} R_\eta)(g \otimes_{\tilde{\tau}} \eta_1) : \eta,\ \eta_1$ are $\tilde{\tau}$-bounded, $T_1 \in \mathcal{R}(\mathcal{G})'$ and $g \in L^2(\mathcal{G}, \nu) \}$ is total in ${}_{\mathcal{Z}}L^2(\mathcal{G}, \nu) \otimes_{\tilde{\tau}} {}_{\mathcal{Z}}L^2(\mathcal{G}', \tau_1') \otimes_{\tilde{\tau}} {}_{\mathcal{Z}}L^2(\mathcal{G}', \tau_1')$, we conclude that

$$\Omega(f) = (\iota *_{\mathcal{Z}} \Gamma')\big(\delta\big(L(f)\big)\big).$$

This proves our assertion. Q.E.D.

As a combination of the two previous lemmas, one has

$$(\delta *_{\mathcal{Z}} \iota) \circ \delta(a) = (\iota *_{\mathcal{Z}} \Gamma') \circ \delta(a), \quad (a \in \mathcal{R}(\mathcal{G})).$$

Therefore, we finally obtain

Theorem 8.3. *The system* $(\mathcal{G}',\ \mathcal{R}(\mathcal{G}),\ \delta,\ L^2(\mathcal{G}, \nu))$ *is a coaction of* $\mathcal{G}'$*.*

Remark 8.4. If the groupoid $\mathcal{G}$ happens to be principal, then the derived groupoid $\mathcal{G}'$ coincides with $\mathcal{G}$ itself, so that we have $\mathcal{R}(\mathcal{G}') = \mathcal{R}(\mathcal{G})$. In this case, it can be shown that the above defined coaction δ is nothing but the coproduct Γ of $\mathcal{R}(\mathcal{G})$. Hence interesting things may happen if $\mathcal{G}$ is not principal.

References

[**B**] R. Bowen, *Anosov foliations are hyperfinte*, Ann. of Math. **106** (1977) 549–565.

[**CLN**] C. Camacho and A. Lins Neto, *Geometric theory of foliations*, Boston, Birkhauser, 1985.

[**C1**] A. Connes, *Une classification des facteurs de type III*, Ann. Sci. École Norm. Sup. **6** (1973) 133–252.

[**C2**] ———, *The von Neumann algebra of a foliation*, Lecture Notes in Phys., Springer-Verlag **80** (1978) 145–151.

[**C3**] ———, *Sur la théorie non commutative de l'intégration*, Lecture Notes in Math., Springer-Verlag **725** (1979) 19–143.

[**C4**] ———, *On the spatial theory of von Neumann algebras*, J. Functional Analysis **35** (1980) 153–164.

[**CT**] A. Connes and M. Takesaki, *The flow of weights on factors of type* III, Tohoku Math. J. **29** (1977) 473–575.

[**D**] J. Dixmier, *Von Neumann algebras*, North-Holland. 1981.

[**E**] E. G. Effros, *Global sturcture in von Neumann algebras*, Trans. Amer. Math. Soc. **121** (1966) 434–454.

[**ES**] M. Enock and J-M. Schwartz, *Une dualité dans les algébres de von Neumann*, Bull. Soc. France, Suppl. **44** (1975) 1–144.

[**F**] J. M. G. Fell, *An extension of Mackey's method to Banach $*$-algebraic bundles*, Memoirs Amer. Math. Soc. **90** (1969).

[**FMI**] J. Feldman and C. C. Moore, *Ergodic equivalence relations, cohomology and* *von Neumann algebras I*, Trans. Amer. Math. Soc. **234** (1977) 289–324.

[**FMII**] ———, *Ergodic equivalence relations, cohomology and von Neumann algebras II*, Trans. Amer. Math. Soc. **234** (1977) 325–359.

[**Haa**] U. Haagerup, *The standard form of von Neumann algebras*, Math. Scand. **37** (1975) 271–283.

[**Ha1**] P. Hahn, *Haar measure for measure groupoids*, Trans. Amer. Math. Soc. **242** (1978) 1–33.

[**Ha2**] ———, *The regular representations of measure groupoids*, Trans. Amer. Math. Soc. **242** (1978) 34–72.

[**HO**] T. Hamachi and M. Osikawa, *Ergodic groups of automorphisms and Krieger's theorem*, Keio Univ., Seminar on Math. Sci. No.**3** (1981)

[**JT**] V. F. R. Jones and M. Takesaki, *Actions of compact abelian groups on semifinite injective factors*, Acta Math. **153** (1984) 213–258.

[**K**] D. Kastler, *On A. Connes' noncommutative integration theory*, Comm. Math. Phys. **85** (1982) 99–120.

[**Ko**] H. Kosaki, *Canonical L^p-spaces associated with an arbitrary abstract von Neumann algebra*, Dissertation at UCLA (1980).

[**M**] G. W. Mackey, *Ergodic theory and virtual groups*, Math. Ann. **166** (1966) 187–207.

[**Ma**] T. Masuda, *Groupoid dynamical systems and crossed product I–The case of W^*-systems*, Publ. R.I.M.S. Kyoto Univ. **20** (1984) 929–957.

[**MS**] C. C. Moore and C. Schochet, *Global analysis on foliated spaces*, MSRI Publ.

9 Springer-Verlag. 1988.

[**NT**] Y. Nakagami and M. Takesaki, *Duality for crossed products of von Neumann algebras*, Lecture Notes in Math. Springer-Verlag **731** (1979)

[**Ra1**] A. Ramsay, *Virtual groups and group actions*, Adv. in Math. **6** (1971) 253–322.

[**Ra2**] ———, *Boolean duals of virtual groups*, J. Functional Analysis **15** (1974) 56–101.

[**Ra3**] ———, *Topologies on measured groupoids*, J. Functional Analysis **47** (1982) 314–343.

[**R**] J. Renault, *A groupoid approach to C^*-algebras*, Lecture Notes in Math. Springer-Verlag **793** (1980)

[**Sa1**] J. L. Sauvageot, *Produits tensoriels de $\mathcal{Z}$-modules*, Publ. Univ. P. & M. Curie n° 23 (1980)

[**Sa2**] ———, *Produits tensoriels de $\mathcal{Z}$- modules et application*, Lecture Notes in Math. Springer-Verlag **1132** (1983) 468–485.

[**Sa3**] ———, *Sur le produit tensoriel relatif d'espaces de Hilbert*, J. Operator Theory **9** (1983) 237–252.

[**Se**] C. Series, *The Poincaré flow of a foliation*, Amer. J. Math. **102** (1980) 93–128.

[**ST**] C. Sutherland and M. Takesaki, *Actions of amenable groups and groupoids on semifinte injective von Neumann algebras*, R.I.M.S. Kyoto Univ. **21** (1985) 1087–1120.

[**T1**] M. Takesaki, *Duality and von Neumann algebras*, Lecture Notes in Math. Springer-Verlag **247** (1972) 666–779.

[**T2**] ———, *Duality for crossed products and the structure of von Neumann algebras* *of type III*, Acta Math. **131** (1973) 249–310.

[**T3**] ———, *Theory of operator algebras I*, Springer-Verlag 1979.

[**T4**] ———, *The structure of operator algebras* (in Japanese), Iwanami-shoten. 1983.

[**T4**] ———, *A classification theory based on groupoids* in *Geometric methods in operator algebras*, Pitman Res. Notes in Math. Vol. **123** (1986) 400–410.

[**Y1**] T. Yamanouchi, *Duality for actions and coactions of measured groupoids on von Neumann algebras*, Dissertation at UCLA (1990).

[**Y2**] ———, *Duality for actions and coactions of groupoids on von Neuamnn algebras*, Submitted for publication.

[**Y3**] ———, *Duality for generalized Kac algebras and a characterization of finite groupoid algebras*, Submitted for publication.

[**Y4**] ———, *Crossed products by groupoid actions and their smooth flows of weights*, Preprint.

[**Z**] R. J. Zimmer, *Ergodic theory and semisimple groups*, Monograph in Math. Boston, Birkhauser. 1984.

Department of Mathematics and Computer Science

University College of Swansea

Singleton Park, Swansea

SA2 8PP United Kingdom

Editorial Information

To be published in the *Memoirs*, a paper must be correct, new, nontrivial, and significant. Further, it must be well written and of interest to a substantial number of mathematicians. Piecemeal results, such as an inconclusive step toward an unproved major theorem or a minor variation on a known result, are in general not acceptable for publication. *Transactions* Editors shall solicit and encourage publication of worthy papers. Papers appearing in *Memoirs* are generally longer than those appearing in *Transactions* with which it shares an editorial committee.

As of November 1, 1992, the backlog for this journal was approximately 9 volumes. This estimate is the result of dividing the number of manuscripts for this journal in the Providence office that have not yet gone to the printer on the above date by the average number of monographs per volume over the previous twelve months. (There are 6 volumes per year, each containing about 3 or 4 numbers.)

A Copyright Transfer Agreement is required before a paper will be published in this journal. By submitting a paper to this journal, authors certify that the manuscript has not been submitted to nor is it under consideration for publication by another journal, conference proceedings, or similar publication.

Information for Authors

Memoirs are printed by photo-offset from camera copy fully prepared by the author. This means that the finished book will look exactly like the copy submitted.

The paper must contain a *descriptive title* and an *abstract* that summarizes the article in language suitable for workers in the general field (algebra, analysis, etc.). The *descriptive title* should be short, but informative; useless or vague phrases such as "some remarks about" or "concerning" should be avoided. The *abstract* should be at least one complete sentence, and at most 300 words. Included with the footnotes to the paper, there should be the 1991 *Mathematics Subject Classification* representing the primary and secondary subjects of the article. This may be followed by a list of *key words and phrases* describing the subject matter of the article and taken from it. A list of the numbers may be found in the annual index of *Mathematical Reviews*, published with the December issue starting in 1990, as well as from the electronic service e-MATH [**telnet e-MATH.ams.org** (or **telnet 130.44.1.100**). Login and password are **e-math**]. For journal abbreviations used in bibliographies, see the list of serials in the latest *Mathematical Reviews* annual index. When the manuscript is submitted, authors should supply the editor with electronic addresses if available. These will be printed after the postal address at the end of each article.

Electronically-prepared manuscripts. The AMS encourages submission of electronically-prepared manuscripts in $\mathcal{A}\mathcal{M}\mathcal{S}$-TeX or $\mathcal{A}\mathcal{M}\mathcal{S}$-LaTeX. To this end, the Society has prepared "preprint" style files, specifically the amsppt style of $\mathcal{A}\mathcal{M}\mathcal{S}$-TeX and the amsart style of $\mathcal{A}\mathcal{M}\mathcal{S}$-LaTeX, which will simplify the work of authors and of the production staff. Those authors who make use of these style files from the beginning of the writing process will further reduce their own effort.

Guidelines for Preparing Electronic Manuscripts provide additional assistance and are available for use with either $\mathcal{A}_{\mathcal{M}}\mathcal{S}$-TEX or $\mathcal{A}_{\mathcal{M}}\mathcal{S}$-LATEX. Authors with FTP access may obtain these *Guidelines* from the Society's Internet node e-MATH.ams.org (130.44.1.100). For those without FTP access they can be obtained free of charge from the e-mail address guide-elec@math.ams.org (Internet) or from the Publications Department, P. O. Box 6248, Providence, RI 02940-6248. When requesting *Guidelines* please specify which version you want.

Electronic manuscripts should be sent to the Providence office only after the paper has been accepted for publication. Please send electronically prepared manuscript files via e-mail to pub-submit@math.ams.org (Internet) or on diskettes to the Publications Department address listed above. When submitting electronic manuscripts please be sure to include a message indicating in which publication the paper has been accepted.

For papers not prepared electronically, model paper may be obtained free of charge from the Editorial Department at the address below.

Two copies of the paper should be sent directly to the appropriate Editor and the author should keep one copy. At that time authors should indicate if the paper has been prepared using $\mathcal{A}_{\mathcal{M}}\mathcal{S}$-TEX or $\mathcal{A}_{\mathcal{M}}\mathcal{S}$-LATEX. The *Guide for Authors of Memoirs* gives detailed information on preparing papers for *Memoirs* and may be obtained free of charge from AMS, Editorial Department, P.O. Box 6248, Providence, RI 02940-6248. The *Manual for Authors of Mathematical Papers* should be consulted for symbols and style conventions. The *Manual* may be obtained free of charge from the e-mail address cust-serv@math.ams.org or from the Customer Services Department, at the address above.

Any inquiries concerning a paper that has been accepted for publication should be sent directly to the Editorial Department, American Mathematical Society, P. O. Box 6248, Providence, RI 02940-6248.

Recent Titles in This Series

(Continued from the front of this publication)

452 **Brian S. Thomson,** Derivates of interval functions, 1991

451 **Rolf Schön,** Effective algebraic topology, 1991

450 **Ernst Dieterich,** Solution of a non-domestic tame classification problem from integral representation theory of finite groups ($\Lambda = RC_3, v(3) = 4$), 1991

449 **Michael Slack,** A classification theorem for homotopy commutative H-spaces with finitely generated mod 2 cohomology rings, 1991

448 **Norman Levenberg and Hiroshi Yamaguchi,** The metric induced by the Robin function, 1991

447 **Joseph Zaks,** No nine neighborly tetrahedra exist, 1991

446 **Gary R. Lawlor,** A sufficient criterion for a cone to be area-minimizing, 1991

445 **S. Argyros, M. Lambrou, and W. E. Longstaff,** Atomic Boolean subspace lattices and applications to the theory of bases, 1991

444 **Haruo Tsukada,** String path integral realization of vertex operator algebras, 1991

443 **D. J. Benson and F. R. Cohen,** Mapping class groups of low genus and their cohomology, 1991

442 **Rodolfo H. Torres,** Boundedness results for operators with singular kernels on distribution spaces, 1991

441 **Gary M. Seitz,** Maximal subgroups of exceptional algebraic groups, 1991

440 **Bjorn Jawerth and Mario Milman,** Extrapolation theory with applications, 1991

439 **Brian Parshall and Jian-pan Wang,** Quantum linear groups, 1991

438 **Angelo Felice Lopez,** Noether-Lefschetz theory and the Picard group of projective surfaces, 1991

437 **Dennis A. Hejhal,** Regular b-groups, degenerating Riemann surfaces, and spectral theory, 1990

436 **J. E. Marsden, R. Montgomery, and T. Ratiu,** Reduction, symmetry, and phase mechanics, 1990

435 **Aloys Krieg,** Hecke algebras, 1990

434 **François Treves,** Homotopy formulas in the tangential Cauchy-Riemann complex, 1990

433 **Boris Youssin,** Newton polyhedra without coordinates Newton polyhedra of ideals, 1990

432 **M. W. Liebeck, C. E. Praeger, and J. Saxl,** The maximal factorizations of the finite simple groups and their automorphism groups, 1990

431 **Thomas G. Goodwillie,** A multiple disjunction lemma for smooth concordance embeddings, 1990

430 **G. M. Benkart, D. J. Britten, and F. W. Lemire,** Stability in modules for classical Lie algebras: A constructive approach, 1990

429 **Etsuko Bannai,** Positive definite unimodular lattices with trivial automorphism groups, 1990

428 **Loren N. Argabright and Jesús Gil de Lamadrid,** Almost periodic measures, 1990

427 **Tim D. Cochran,** Derivatives of links: Milnor's concordance invariants and Massey's products, 1990

426 **Jan Mycielski, Pavel Pudlák, and Alan S. Stern,** A lattice of chapters of mathematics: Interpretations between theorems, 1990

425 **Wiesław Pawłucki,** Points de Nash des ensembles sous-analytiques, 1990

424 **Yasuo Watatani,** Index for C^*-subalgebras, 1990

423 **Shek-Tung Wong,** The meromorphic continuation and functional equations of cuspidal Eisenstein series for maximal cuspidal subgroups, 1990

(See the AMS catalog for earlier titles)